Fundamentals of Solar Radiation

Fundamentals of Solar Radiation

Lucien Wald

MINES PARISTECH, FRANCE

CRC Press
Taylor & Francis Group
Boca Raton London New York

CRC Press is an imprint of the
Taylor & Francis Group, an **informa** business

Cover image: *Hêtre tortillard* (Fagus sylvatica Tortuosa), Chapel of Sainte-Corneille, Forest of Compiègne, France. Photo credit: Roseline Adde.

Translated from the French *Introduction au Rayonnement Solaire*, first published by Presses des Mines, Paris, 2020.

CRC Press/Balkema is an imprint of the Taylor & Francis Group, an informa business

© 2021 Taylor & Francis Group, London, UK

Typeset by codeMantra

Library of Congress Cataloging-in-Publication Data
Names: Wald, Lucien, author.
Title: Fundamentals of solar radiation / Lucien Wald, MINES ParisTech, France.
Description: Boca Raton : CRC Press, [2021] | Includes index.
Subjects: LCSH: Solar radiation.
Classification: LCC QB531 .W35 2021 (print) | LCC QB531 (ebook) | DDC 523.7/2–dc23
LC record available at https://lccn.loc.gov/2020050520
LC ebook record available at https://lccn.loc.gov/2020050521

Published by: CRC Press/Balkema
                Schipholweg 107C, 2316 XC Leiden, The Netherlands
                e-mail: Pub.NL@taylorandfrancis.com
                www.crcpress.com – www.taylorandfrancis.com

ISBN: 978-0-367-72588-4 (Hbk)
ISBN: 978-0-367-72592-1 (Pbk)
ISBN: 978-1-003-15545-4 (eBook)

DOI: 10.1201/9781003155454
https://doi.org/10.1201/9781003155454

Contents

Figures

Tables

Preface

The Sun knew long before me that I would be its lifelong disciple. Harbingers were sprinkled throughout my early life. I was born in Tunisia, a Mediterranean country bathed by the sun, and grew up in France. In my teenage years, I studied in a military college located in the surroundings of Paris, whose buildings housed a boarding school for girls founded two centuries ago by the most-cherished wife of the king Louis XIV of France, also known as the SUN King! There, the Sun trapped me, and 15 years later, I would devote my days to research on solar radiation, in an institute located in the sunny French Riviera.

My debut in solar radiation research was both accidental and odd. I was a PhD student in oceanography and was processing images of the Earth taken by satellites in order to map surface currents in the Mediterranean Sea. I was pushing hard, working day and night, 7 days a week, because I was on a 1-year contract. During that time, I was asked to perform an analysis of the presence of clouds because I also had been looking into atmosphere physics. I did it very reluctantly, and once finished, I quickly returned to my study. By coincidence, this analysis was one of the few seminal works that created a worldwide interest in the use of images taken by satellites for estimating the solar radiation received at ground level. Today, these early works gave birth to algorithms that are used every day in many important applications, as it will be seen further.

A few years later, circumstances led me to get involved in solar radiation research mostly to help in assessing the solar potential for applications in energy production by photovoltaic panels or solar water heaters. By inventing complex but fast algorithms to extract information on the physics of the atmosphere from images taken by satellites, I was to provide accurate modeling of the effects of the atmosphere on solar radiation. I really liked the subject and I brought several innovations in this area. What started by accident has become a life full of very attractive challenges in atmospheric physics and data science.

A book for anyone interested in the sun

In 2000, my team and I created a new monitoring service, called HelioClim, which streams real-time or archived solar radiation data, hence called the SoDa project. When it started, access to such data was difficult and costly, and having this database freely available revolutionized daily practices in numerous companies and other organizations. There were, and still are some 20 years later, a large number of users of this service and those that follow. I have personally met many of them – researchers, academics, teachers, engineers, technicians, or students, working in different fields, from human health to material weathering, – and several were keen to learn the basics of solar radiation, which I was only too happy to provide. I also gave masters- and doctorate-level university courses in different fields in science and engineering, or in continuing education, engineer level. This book is an extension of my pleasure at being able to share. It is for anyone who has questions about solar radiation and is looking for simple answers. Its ambition is to help them by presenting the fundamental elements of the solar radiation received at ground level.

Why should we care about the sun?

Reader, you know that the sun is a star. It is very hot, much hotter than a furnace, and because of this, it radiates a tremendous amount of energy, called solar energy or solar radiation. Only a part of it reaches the Earth and its atmosphere, but it is still a considerable amount. Experts have calculated that at any one time, the Earth intercepts approximately $180 \ 10^6$ GW. This number is also written as $18 \ 10^{13}$ W, or $180,000,000,000,000,000$ W – with 13 zeroes to spare you the counting. Obviously, no one can truly fathom the sheer amount of energy this entails, but some experts say that each minute the Earth receives an amount of energy equivalent to the annual consumption of Europe in electricity! Frankly, even with these examples, such quantities still remain hard to grasp.

In contrast, I fully understand that solar radiation is the main natural source of energy on Earth and by a long way. The other sources are the geothermal heat flux generated from inside the Earth, natural terrestrial radioactivity, and cosmic radiation, which are all negligible compared to solar radiation, since they produce about 0.1 % of the total amount. In other words, solar radiation is the almost unique supplier of energy to the Earth. It has a primary influence on many phenomena, natural or not, on life on Earth as well as on other aspects of human health and society. Whatever we do, the sun is present in our daily lives and has an effect on most of our activities.

An example? Let me begin with climate and weather. Unlike the sun, the outer space surrounding the Earth is very cold, about −270 °C. It absorbs all the energy emitted by the Earth. The climate is determined by the balance between the solar radiation received by the Earth and the energy it loses to outer space. Part of the solar radiation is absorbed by the atmosphere, the continents, and the oceans. Briefly speaking, it is converted into heat and is further transported by the atmosphere and the oceans, from surplus regions, such as the equatorial belt, to deficit areas, such as the poles. This transport of energy takes place at all scales of time and space. In the atmosphere, it is responsible for our weather patterns under the form of convection, evaporation, cloud formation, winds, precipitation, etc.

But beyond climate and weather modeling, how does our understanding of solar radiation help in our day-to-day lives? Let me list a few examples.

Solar radiation brings heat to objects, such as a glazed room, the walls of a building, and an airplane cabin. This results in an increase in the internal temperature, which can be pleasant and provide additional comfort. But, if the radiation is too intense and the temperature too high, and if the object has been poorly designed with regard to the heat balance, the result can be harmful. The interior of a building can become excessively hot and uncomfortable, an outdoor gas tank can explode, and the temperature inside an airplane cockpit can become unbearably high, causing pilot discomfort and increased risk. Hence, solar radiation is an input in design and architecture. For buildings, for instance, the architect must take into account solar radiation to size the windows and other glazing to offer an optimal amount of natural light in each room, while seeking a compromise to maintain good thermal comfort.

In greenhouse tomato crops, where the air temperature is regulated, an accurate estimate of the solar radiation determines the water supply. In vineyards, solar radiation affects the ripening of the grapes directly but also indirectly, through the air temperature, the phenology of the grape, and its sugar, flavoring, and polyphenol content. For example, the solar radiation in Bordeaux vineyards decreases from west to east, with a very clear influence on the zones "d'appellation d'origine contrôlée" and the ripening potential of the grape.

Solar radiation participates in the decomposition of chemical compounds including atmospheric pollutants, what is called photolysis. Photolysis occurs continuously, producing secondary pollutants that have harmful effects on human health and plants. Their appearance is favored by strong sunshine. Hence, solar radiation is an input to pollution models and helps raise pollution alerts, saving lives every year. Some organic polymer materials such as PVC, used in construction, paints, varnishes, and outdoor furniture, degrade under the effect of ultraviolet rays. Their physical or chemical properties are modified, their color changes (e.g., from white to gray), and their surface can present cracks and crazing. To guarantee the properties of their products, manufacturers must know the exact solar radiation in regions where they are used.

Solar radiation also works for energy, replacing carbon-based sources and reducing climate impact. For heating water, the heat transfer fluids in thermal solar panels are heated by solar radiation. The heat thus generated is used for the production of domestic hot water, or for heating in the home. In so-called concentration systems, the radiation is concentrated by lenses or by a very large number of mirrors making the sun rays converge at the same point. The amount of heat produced is very large and sufficient to vaporize a liquid and turn a turbine generating electricity. The complexity of concentrating systems requires great precision in the estimation and short-term forecasting of solar radiation. Another way is to use photovoltaic panels, which are very widespread today in our landscapes and on our homes. These panels simply convert the solar radiation they receive into electricity. The choices of location and photovoltaic technologies, as well as the design of a large-scale photovoltaic farm, are guided by an analysis of the solar radiation received on the ground.

The impacts of solar radiation on health are very numerous – sometimes beneficial, sometimes harmful. Our mood is more or less influenced by the presence of clear sky, the intensity of the radiation, and the daytime. Visible light, the light that our eyes perceive, affects our biological rhythms, the cycle of activities and, more generally, the rhythm of our societies. Also, ultraviolet rays allow the synthesis of vitamin D and indirectly help bones in the fixation of calcium. Other benefits have been seen in other conditions, such as multiple sclerosis, thyroid or breast cancer, and Parkinson's disease, to name just a few of the subjects in the studies I have participated in. On the negative side, exaggerated exposure to the sun, without clothing or protection, can lead to sunstroke, sunburn and burns, actinic keratoses (skin lesions), and skin aging. The most common damage to the skin is cancer. Ultraviolet rays are also dangerous for the eyes. A good knowledge of solar radiation and its variations in time and space can lead to more precise epidemiological studies and, by extension, more effective public health preventive measures.

From fundamentals to measurements

Laying on a beach or in another bucolic place, on a beautiful cloudless day, you have certainly observed that you receive more radiation when the sun is high in the sky than when the sun is near the horizon. The position of the sun in the sky relative to a person on the ground, or a collecting surface, such as an alpine lawn, the facade of a building, or even a photovoltaic panel, plays a major role with regard to the amount of solar radiation that this person or surface receives. This is why the first of the three parts of this book is devoted to the relative geometry between the direction of the sun and an observer on the ground as well as to the solar radiation emitted by the sun and received at the top of the atmosphere.

Chapter 1 deals with the definition of time and the different systems of time. It recalls the geocentric and geographic coordinates that play a major role in solar radiation. Without claiming to be as complete and accurate as a work in astronomy, the orbit of the Earth around the sun and the solar declination are described. The concept of time is introduced and is closely linked to the solar cycle and the rotation of the Earth on itself. The orbit of the Earth around the sun defines the year, the astronomical seasons, and the months. The rotation of the Earth on itself with respect to the sun defines the day, the hour, the minute, and the second. Different time systems, namely mean and true solar times, universal time, and legal time, are defined. Equations are given to calculate the time in these different systems.

Chapter 2 describes the angles used to define the course of the sun over an observer on the ground, namely the solar zenithal angle, the solar azimuth, and the angular elevation of the sun above the horizon, and provides equations to compute them at any place and any time. Several graphs are provided to picture the variations of these angles along the year at various latitudes. The chapter addresses the definition and calculations of times of sunrise and sunset and the duration of the day, or daytime, or daylength. Since the radiation measurements are made over a certain period of time during which the solar angles change, what are the solar angles to be assigned to each measurement? Chapter 2 answers this problem by presenting the concept of effective solar angle, as well as solutions for associating the solar angles with each measurement.

Chapter 3 deals with solar radiation received by a surface at the top of the atmosphere, also known as the extraterrestrial radiation. Solar radiation varies over time due to changes in the position of the Earth relative to the sun and, to a lesser extent, variations in solar activity. Radiance, irradiance, irradiation, radiant exposure, and sunshine duration are defined. Equations are given to calculate the solar radiation received on a horizontal or inclined surface located at the top of the atmosphere. Any object whose surface temperature is greater than 0 K (–273 °C) emits radiation. This is the case of the sun whose surface temperature is extremely high (around 5780 K or about 5500 °C) and which radiates over a range of wavelengths, between 200 and 4000 nm. This chapter also describes the spectral distribution of the extraterrestrial radiation which is not uniform over this range. Half of the received power lies in the visible range, the rest being in the ultraviolet and in the near and middle infrared.

Only part of the extraterrestrial radiation reaches the ground, the rest being mainly absorbed or backscattered to the outer space by the constituents of the atmosphere. This proportion is about 70 %–80 % if the sky is clear, i.e., cloudless, and decreases in the presence of clouds. The second part of this book addresses how the solar radiation incident at the top of the atmosphere is attenuated and modified in its downward path to the ground.

Still lying on the beach, you probably have observed that the appearance of clouds, their coverage, and opacity vary over time and according to geographic location. Clouds of low optical thickness, like cirrus clouds, allow a significant amount of radiation to pass through. Large optical clouds scatter and reflect radiation strongly in all directions and can cause noticeable darkness for the ground observer. The interactions between radiation and the medium it crosses are gathered under the term of radiative transfer.

Chapter 4 deals with the radiative transfer in the atmosphere, that is to say, the interactions between solar radiation and the atmosphere and its constituents. It introduces the phenomena of absorption and scattering which play a prominent role. It details the absorption of radiation by atmospheric gases in certain wavelengths, as well as the scattering of radiation by molecules in the air. Next, the effects of aerosols that are airborne particles such as soot or grains of sand are presented, followed by a discussion on the role of clouds, which are the main attenuators of solar radiation. Several diagrams illustrate the different paths that the sun rays can follow due to the scattering into the atmosphere. A summary of the contributions of the constituents of the atmosphere to the depletion of solar radiation is provided.

Part of the radiation reaching the ground is reflected, part of which is backscattered toward the ground. This is addressed in Chapter 5. Several quantities, namely the reflection factor, the reflectance, the bidirectional reflectance distribution function, and the albedo, are defined that describe how the sun rays are reflected by the ground. Several examples of the spectral distribution of the albedo are given.

Both Chapters 6 and 7 combine the elements of the preceding chapters to describe the solar radiation received on the ground by a horizontal or inclined collector plane, such as a natural slope or a rooftop. Chapter 6 deals with the total radiation, while Chapter 7 is dedicated to the spectral distribution.

The solar radiation received at ground level by a horizontal surface has a direct component and a diffuse component. There is also a reflected component due to the reflection of radiation from the surrounding ground, if the receiving surface is tilted or in case of relief. The sum of these components is called global radiation. Chapter 6 defines these components and presents other quantities such as the direct fraction, the diffuse fraction, the clearness index, or the clear-sky index. Clouds deplete the radiation in a prominent way in front of the other attenuators. They exhibit significant variations in time and space. Examples are given to illustrate the influence of these variations on the solar radiation at ground level at different time and space scales. The most influential variables on the solar radiation reaching the ground, such as date and geographic location, or constituents of the clear and cloudy atmosphere, are listed.

Chapter 7 deals with the spectral distribution of the solar radiation received on the ground and its direct and diffuse components. It is difficult to give accurate values of the spectral distribution as it is done in the previous chapters, in view of the wide variety of atmospheric situations and conditions, as discussed in Chapter 6. This is why I favored an illustrative approach, including typical cases carried out by numerical simulation, with the essential aim of making the influence of the atmosphere understood. This chapter highlights the complexity of calculating solar radiation at the ground level under changing sky conditions. It presents a simplification of this calculation, which is of great practical significance.

Solar radiation, like all other meteorological variables and so many others, exhibits fluctuations in time and space, of different durations and sizes. Chapter 8 addresses the variability of the radiation and gives some elements to understand and describe this variability, such as time and space scales. It introduces the random nature of radiation and the concept of ergodicity that allows the comparison between temporal and spatial statistics. The chapter illustrates the influence of this variability on the measurement acquisition processes, the digital processing of measurements, and other methods for estimating solar radiation at ground. Measurement and sampling are defined. The sampling theorem emphasizes the importance of the latter in relation to the properties of solar radiation estimated from the measurements.

The third part deals with direct or indirect measurements of the solar radiation received on the ground over a given integration time (minute, hour, day, or month), whether for total radiation or for radiation in a spectral range such as ultraviolet (UV), or daylight, or photosynthetically active radiation (PAR). It also explains how to check the plausibility of the measurements.

Chapter 9 presents the most common instruments for measuring radiation at ground. Several examples illustrate the heterogeneity of the distribution of radiation measurement stations in the world, as well as the difficulty of obtaining complete time series, that is to say, with little missing data. The averaging of a time series of radiation measurements, for example, the monthly or annual average of the irradiance, is a common operation. The chapter presents first the usual mathematical operations for this calculation and then the recommendations specific to the World Meteorological Organization. Finally, it discusses how to perform such a calculation in the absence of measurements in time series, i.e., in the presence of gaps.

To overcome the problem of incomplete coverage in both space and time by stations, Earth observation instruments, other than dedicated ground-based ones, are used. They include numerical weather models, meteorological analyses or reanalyses, or readings from instruments on board satellites, including multispectral imagers. Chapter 10 presents these means and shows how they can be combined with each other and with measuring ground stations. It also describes the concept of interpolation as some experts and practitioners prefer to perform spatial interpolations of measurements from stations close to their site of interest. Chapter 10 reconciles the variability seen in Chapter 8, the underlying assumptions and the uncertainty resulting from spatial interpolation between stations. Finally, Chapter 10 exposes the concept and principles of empirical relationships to estimate the radiation in a given spectral range from measurements of total radiation, or to estimate the radiation, whether total or in a given spectral range, from measurements of other meteorological variables such as sunshine duration, air temperature at surface, cloud cover, or cloudiness. It does not go into detail as many relationships of this type have been proposed and still are by researchers.

Measurements of an instrument are only of value if they are of quality. Measuring radiation is quite complex, as well as monitoring the quality of the measurements. Suppliers do not always give a quality estimate along with the radiation measurements.

However, knowing their quality is an important element when acquiring measurements from third parties. The purchaser is often obliged to carry out a measurement check, an operation called checking the plausibility of the measurements. This verifies that the measurement is probably correct, that it is not too large or too small compared to what can be expected under the conditions encountered. This plausibility check is the subject of the two last chapters. Chapter 11 presents the objective of the check and illustrates the need for such check. The different elements to check and how to check them are detailed. Chapter 12 addresses visual and automated procedures for verifying measurements. It is not intended to present all possible techniques. It constitutes a sort of vade mecum, a set of simple rules and procedures which allow the plausibility of the measurements to be checked quite easily.

What about climate change? Does this book forget the subject? No significant change in the radiation emitted by the sun is expected in the coming decades. However, changes in climate and associated variables have been observed for several decades, which are mainly due to increases in the concentrations of certain gases in the atmosphere, which lead, among other phenomena, to an increase in air temperature at the surface of the earth.

These changes in air temperature are well described and well modeled for the decades to come. Thanks to the work of scientists and their numerical models, several climate projections are available based on emission scenarios for these gases and aerosols, projections which go in the same direction, although not identical. These changes in atmospheric gas concentrations and air temperatures have a significant impact on radiation. In fact, the solar radiation at ground level results from intertwined phenomena, each interacting with the others. Changes in one cause changes in others and in feedback changes in the first and so on. For example, (1) warming oceans cause changes in atmospheric circulation and therefore changes in the spatial distribution of cloud cover, the structure of clouds and their optical depth, and consequently, the effects of the atmosphere on the solar radiation. The solar radiation at surface may locally be increased or decreased with possible impacts on the production of solar farms, or agricultural productivity. (2) The warming of air at surface causes changes in the albedo of the ground, which affects the reflected part of the radiation. Areas previously covered with snow all year round may no longer be covered, and the reflected part will decrease, which will lead to a decrease in the natural light. In Greenland, the decrease in snow cover led to a decrease in the natural lighting of the roads and an increase in the electric consumption for the latter. (3) The increase in the concentration of certain gases in the atmosphere increases the extinction of the direct component that is the part of the solar radiation that can be concentrated and therefore affects strongly the concentration systems for producing heat or electricity. It may also change the spectral distribution of the radiation with possible impacts on plants.

These few examples are not exhaustive. The interactions between the phenomena are currently modeled, but I think that the results are not all sufficiently accurate for solar radiation. For example, the available climate projections for the next decades for Europe exhibit fairly similar 10-year averages. These averages are fairly close to those observed in the past, meaning that changes in solar radiation will be small on average though many other variables will change. However, the projections show strong variations from 1 year to the next and also great disparities from one projection to another. Why are the results so different for solar radiation between projections and less different for other meteorological variables such as air temperature at surface? The models are certainly correct by themselves, but observations of many variables are currently lacking to reduce the uncertainties and to enrich the models. At the time of writing this book, I believe the uncertainties of projections of solar radiation are large. Although I have often been asked questions about this subject, I think it is still too new to be taught to non-specialist readers and to be the subject of a chapter in this book.

Reader, allow me one last piece of advice. Solar radiation is a physical variable occurring in many fields, as mentioned at the beginning of this introductory chapter. Since each field has its own lingo and habits, I suggest that you be very precise in the use of definitions, descriptions of the quantities used and their use. This could avoid unnecessary confusion, misunderstandings, and disappointments.

In conclusion, I emphasize that this book is only an introduction to solar radiation and should help to know and understand the physical mechanisms involved. It cannot replace the regularly published studies, presenting recent scientific advances. With this book, I present a solid physical foundation on the topic of solar radiation. For this purpose, the book includes many examples and numerous illustrations, as well as some simple methods. It also gives some simple but fairly accurate equations to calculate the various elements covered and to reproduce the figures and graphs.

I hope that, with this knowledge, you will be equipped to better understand the most recent scientific results and develop your own understanding. Like so many other authors of scientific works, my ultimate wish is that you enjoy every moment of your reading and that after having finished it, you will exclaim, not εὐρηκα (Eureka) like Archimedes, but το κατάλαβα (to katálava), that is: I understood! But why in Greek? Well, to pay homage to the Greek philosopher and scientist Aristotle,[1] who wrote the first known book on meteorology, over 2300 years ago!

---

1 Born in 384 B.C., died in 322 B.C.

Acknowledgments

Thanks to Roseline for her understanding and her wonderful daily support. Daily life is not always easy when your other half is lost in thought. Although she denies it, Silvia Dekorsy, my French editor, played an important but distant role in the realization of this project. It is easy to imagine yourself as the author of a book. It is another matter of writing it. After promising her, I had to do it even though there was no obligation. Silvia was my conscience that brought me discreetly and regularly to the object of my promise, for which I thank her deeply. The quality of her work as an editor of my works in French inspired and encouraged me to take even more care in this writing.

Thanks to Philippe Blanc for his scientific and technical help in writing this book. I would have liked to write this book with him. It could not be done. Without doubt, Philippe knew much more than me, the time, the effort, and the commitment necessary to write such a book, which more is in a foreign language. Regardless, his mind hovered unconsciously over these pages. Thanks to Mireille Lefèvre for her help in simulating radiative transfer in the atmosphere. Her knowledge in using the libRadtran simulator was invaluable to me. I have a special thought for Bella Espinar who convinced me in 2006 to write some introductory pages on solar radiation in the energy section for the UNESCO's Encyclopedia of Life Support Systems (EOLSS). Neither she nor I, we suspected that this would lead to several successive writings of which this book is the ultimate culmination.

Mesdames Mireille Lefèvre, Christelle Rigollier and Sylvie Wald, Messieurs Sylvain Cros, Nicolas Fichaux, Thierry Ranchin, Lionel Ménard, and Fabien Wald did me the great service of rereading the first version of this book and helped me with their comments to improve this writing. I express my gratitude to them here.

Some graphs in this book have been produced using the Web services available on the SoDa Service. May the team of Transvalor, which takes care of it, be thanked here for this contribution to the common good. I thank those who believed that I was an expert in the field, whether they were high school students, students, researchers, engineers, or others, and in particular, those who were my students. As I would have been ashamed to disappoint them, I had to constantly consolidate and improve my knowledge of the subject in order to better meet their expectations. Finally, I would like to thank all the other experts who willingly shared their own knowledge with me, in particular via Wikipedia, which is a great source of knowledge, provided you keep a critical eye.

Chapter 1

The definition of time and different time systems

At a glance

The Earth follows an elliptical, quasi-circular, orbit around the sun counter-clockwise. The plane containing the orbit is called the plane of the ecliptic. An orbit takes a year of about 365.25 days to complete, with 1 day of 24 h, or 86,400 s. The distance between the sun and the Earth is an annual average of 1 astronomical unit (1 au = 1.496 10^8 km). It varies about 3 % over the year.

The equatorial plane of the Earth is inclined relative to the plane of the ecliptic. The solar declination is the angle between the equatorial plane and the line linking the sun to the Earth. It varies over the year between $-23.45°$ and $+23.45°$.

Time is closely related to the solar cycle. The orbit of the Earth around the sun and the solar declination define the year, the astronomical seasons, and the months. The equinoxes of March and September are the points of the orbit where the solar declination is zero. The solstices of June and December are the points where the declination reaches its extremes. Boreal winter (in the northern hemisphere), respectively austral summer (in the southern hemisphere), begins on December solstice, i.e., between 20 and 22 December. Boreal summer, respectively austral winter, begins on June solstice, i.e., between 20 and 22 June. The boreal springs and autumns, respectively austral autumns and springs, begin on equinoxes of March (between 19 and 21) and September (between 21 and 24).

The rotation of the Earth on itself with respect to the sun defines a mean solar day, divided into 24 h of 60 min each, or 3600 s each. The mean solar time is such that as an average over the year, the sun is at its highest when the mean solar time is equal to 12:00. It depends upon the longitude; a difference of 1 h corresponds to a difference of 15° in longitude. The true solar time is such that the sun is at its highest when the solar time is equal to 12:00 each day. The difference between the true and mean solar times is a function of the number of the day in the year. The maximum is equal to 0.276 h ($\approx$17 min) and is reached around 31 October. The minimum is -0.242 h ($\approx$15 min), reached around 13 February.

The international reference for legal time is the coordinated universal time (UTC). It is very close to the mean solar time at longitude 0°, less than 1 s. The legal time is that used in a country. It is usually equal to the UTC plus or minus a shift in hours. It is also called civil time, or local time. Relationships link the legal, UTC, mean, and true solar times.

The movement of the sun across the sky, also called the course of the sun in the sky, plays a major role in the amount of solar radiation received at ground level and at the top of the atmosphere. Time is closely related to the solar cycle. The orbit of the Earth around the sun defines the year, the astronomical seasons, and the months. The rotation of the Earth on itself with respect to the sun defines the day, then the hours, the minutes, and the seconds. This chapter deals with the definition of time and the different systems of time.

The legal time that is daily used is related to the solar time and depends on the longitude of the point of interest. As a consequence, this chapter begins with a reminder of the geocentric and geographic coordinates. They describe the position of a point at the surface of the Earth as well as at the top of the atmosphere. Then, the orbit of the Earth around the sun and the solar declination are described. The concept of time is introduced, and different time systems, namely mean and true solar times, universal time, and legal time, are defined.

In this book, I use the Gregorian calendar to indicate dates. This calendar is the one in use in many countries of the world. The time system used in this calendar is closely linked to the position of the Earth relative to the sun.

1.1 Reminder – geocentric and geographic coordinates – angle of incidence

Latitudes and longitudes are angles to describe the position of a point on a perfect or quasi-perfect sphere. The usual notations for latitudes and longitudes are Φ and λ.

The latitude Φ is the angle formed by the equatorial plane and the local vertical. Points of the same latitude form a circle called a parallel. Latitudes are sometimes called parallels. In the ISO 19115 standard[1] used in this book, latitudes are usually expressed in degrees. They are counted positively in the northern hemisphere, from 0° (equator) to 90° (or $\pi/2$, North Pole), and negatively in the southern hemisphere, from 0° to −90° (or −$\pi/2$, South Pole).

Longitudes λ cut the sphere into equal parts, like melon slices. They are defined by planes normal to the equatorial plane containing the center of the Earth. They are also called meridians. The plane passing through the observatory located in the city of Greenwich in the United Kingdom was arbitrarily chosen as the reference and bears the longitude 0°. The longitude of a place is the angle formed by the plane passing through this place and by the reference plane. Longitudes are not parallel since they join at the poles. In the ISO standard, longitudes are counted positively east of the 0° meridian, from 0° to 180° (or π), and negatively west of this meridian, from 0° to −180° (or −π).

As a first approximation, the Earth is a perfect sphere. Latitudes and longitudes defined on this sphere are called geocentric coordinates and are noted Φ_c and λ_c, respectively. Actually, the Earth is not a perfect sphere. It is slightly flattened at the poles, or in other words, it is slightly widened at the equator. The distance R_{pole} from

---

1 ISO 19115-1:2014, 2014. *Geographic Information – Metadata – Part 1: Fundamentals*, 167 p, International Standards Organization, Geneva, Switzerland.

the center of the Earth to a pole is about 6357 km. It is slightly less than the Earth radius at the equator $R_{equator}$ of about 6378 km.

Geographic coordinates Φ and λ are associated with a geodetic system, which is a mathematical expression aimed at getting as close as possible to the gravity field of the Earth. The most widely used geodetic system in the world today is the World Geodetic System 84 (WGS84). It is the basis of the Global Positioning System (GPS) frequently used to know the geographic coordinates of a place. This geodetic system differs from the sphere since it takes into account in particular the flattening at the poles. This is why the geocentric and geographic coordinates differ. The difference must be accounted for when calculating the angles defining the position of the sun relative to a ground observer.

Geocentric λ_c and geographic λ longitudes are equal within a very small error. This is not the case for latitudes for which the difference may be noticeable. The following equation relates Φ_c to Φ:

$$\tan(\Phi_c) = \left(R_{pole} / R_{equator} \right)^2 \tan(\Phi) \tag{1.1}$$

Since the ratio $(R_{pole}/R_{equator})$ is less than 1, $\tan(\Phi_c)$ is less than $\tan(\Phi)$. The difference $(\Phi - \Phi_c)$ is zero at the poles and the equator. It reaches a maximum of 0.0033 rad (0.19°) at the geographic latitudes $\pi/4$ (45°) and $-\pi/4$ (−45°).

As a first approximation, the perimeter P of the Earth along a longitude is:

$$P = 2\pi \left[\left(R_{pole} + R_{equator} \right) / 2 \right] \tag{1.2}$$

Hence, the average distance at the surface of the globe equivalent to 1° in latitude is equal to $P/360°$, or 111 km. Consequently, the greatest error of 0.19° on latitude is equivalent to a maximum error of 21 km ($\approx$0.19 × 111 km) in the north–south direction. If such an error is not acceptable, the latitude Φ must be replaced by the geocentric latitude Φ_c in all equations.

Before leaving this subject of coordinates, I want to treat one additional point. I recall that since the latitudes and longitudes are angles defined on a sphere, perfect or almost perfect, the geographic coordinates (latitude, longitude) of a place located on the ground are identical to those of the point located on the same vertical but at the top of the atmosphere. This is a result that goes without a doubt for you, sagacious reader, and that I will use continuously in this book. Similarly, a horizontal surface on the ground is parallel to the horizontal surface located at the same latitudes and longitudes, but at the top of the atmosphere.

The angle of incidence is the angle describing the inclination of the incident ray relative to the receiving plane. Let me suppose an opaque collector plane, receiving parallel light rays. This is the case, for example, of a plane located at the top of the atmosphere, exposed to the rays of the sun. The angle of incidence θ is defined as the angle between the direction of the incident rays and the direction perpendicular to the plane, as shown in Figure 1.1.

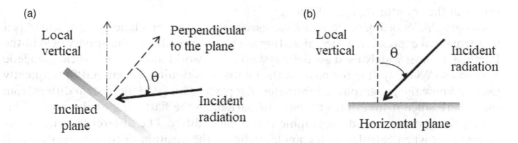

Figure 1.1 Definition of the angle of incidence θ. (a) The general case of an inclined plane; and (b) the case of a horizontal plane.

In geometry, the perpendicular to a surface is also called normal to the surface. The angle of incidence θ is equal to 0 when the incident rays are perpendicular to the plane. This is called normal incidence. Conversely, θ is equal to $\pi/2$ (90°) when the incident rays are parallel to the plane. This is called grazing incidence.

Why should you pay a special attention to the angle of incidence? Why is it so important? This angle plays a preeminent role in the transfer of energy by radiation between the incident radiative energy $Q_{incident}$ and the receiving plane. Indeed, the smaller the angle of incidence of the rays, the greater the radiative energy $Q_{received}$ received by the plane. The following equation holds:

$$Q_{received} = Q_{incident}\cos(\theta) \qquad (1.3)$$

At normal incidence, $\cos(\theta) = 1$, and the received energy is equal to the incident radiative energy. On the contrary, at grazing incidence, the surface intercepted by the rays is zero $(\cos(\theta) = 0)$ and the received energy is zero.

Before leaving this reminding section, I want to treat one additional point regarding angles. I express them in radians or degrees. Regarding the latter, I note them in a decimal or sexagesimal notation. With the increasingly frequent use of software or spreadsheets, it is often necessary to express angles in the decimal system, although it is often preferable to use radians in software, with 1 rad $= (180°/\pi)$. Converting an angle from a sexagesimal format to a decimal format is done by first dividing the number of seconds by 60. The result is added to the number of minutes; the whole is then divided by 60. The result is added to the number of degrees and forms the decimal part. For example:

$$23°27'54'' \rightarrow 23°27' + 54/60 \rightarrow 23°27.9' \rightarrow 23° + 27.9/60 \rightarrow 23.465°$$

Reciprocally, converting an angle from a decimal format to a sexagesimal format is done by first multiplying the decimal part by 60. It gives the number of minutes in decimal format. The decimal part is then multiplied by 60 and yields the number of seconds:

$$23.465° \rightarrow 23° + 0.465 \times 60 \rightarrow 23°27.9' \rightarrow 23°27' + 0.9 \times 60 \rightarrow 23°27'5$$

1.2 Orbit of the Earth around the sun – distance between the sun and the Earth

The Earth follows an elliptical orbit around the sun, as shown schematically in Figure 1.2. In this figure, I have exaggerated the ellipse. The actual eccentricity of the orbit is small (0.01675), which means that the orbit is almost circular. The orbit is flat, and the plane of the orbit is called the plane of the ecliptic, also called the ecliptic in short. The sun is located at one of the focal points of the ellipse and not at its center, which causes variations in the distance between the sun and the Earth during a revolution.

The direction of revolution is counterclockwise, also called trigonometric, or direct (Figure 1.2). The duration of an orbit varies over time over very long periods. This variation can be neglected at the scale of a few centuries. The duration of an orbit is 1 year of 365.25 days.

The distance between the sun and the Earth averaged over a year, r_0, is about 1.496 10^8 km, exactly 149,597,870,700 m according to the decision of the International Astronomical Union in 2012. This distance defines the astronomical unit, abbreviated as au. The distance varies during the year (Figure 1.2). The smallest distance is 0.983 au and is reached between 2 and 5 January. This point of the orbit is called perihelion. The maximum is 1.017 au. It is reached between 3 and 5 July. This point of the orbit is called aphelion. The distance is equal to 1 au twice in the year, between 4 and 5 April and between 4 and 5 October.

Many mathematical expressions for this distance r have been published. They often use truncated series of Fourier coefficients. In this book, more for illustration than for precision, I use the expression proposed by the European Solar Radiation

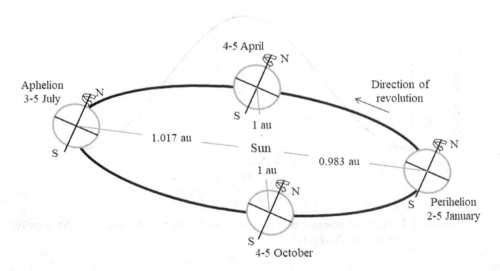

Figure 1.2 Schematic view showing the orbit of the Earth around the sun. The Earth is represented in section with the equatorial plane and its axis of rotation. The distance between the sun and the Earth is given for the two extreme points and the two points for which it is equal to 1 au, with the corresponding dates.

Atlas (ESRA),[2] in which r is a function of the number of the day in the year, num_{day}. num_{day} varies from 1 on 1 January to 365, or 366 in the case of a leap year. It is convenient to define an angle, called the day angle, $angle_{day}$, which is the angle formed by the sun–Earth line for the day num_{day}, and that for 1 January:

$$angle_{day} = num_{day} \, 2\pi/365.2422 \qquad (1.4)$$

Here, $angle_{day}$ is expressed in rad. $angle_{day}$ is close to 0 on 1 January (0.0172). It is equal to π (180°) on 1 July and to 2π (360°) on 31 December. The distance between the sun and the Earth r for the day num_{day}, is given by:

$$\left(r_0/r\right)^2 = 1 + \varepsilon \qquad (1.5)$$

where the correction of relative eccentricity ε is a function of $angle_{day}$:

$$\varepsilon \approx 0.03344\cos\left(angle_{day} - 0.049\right) \qquad (1.6)$$

with sufficient precision in most applications. ε varies between 0.97 and 1.03. The variation of the distance r as a function of the number of the day in the year is shown in Figure 1.3.

The distance r is computed between the barycenters of the sun and of the Earth. In estimating the solar radiation received at the surface, should you take into account the distance between the barycenter of the Earth and the surface, i.e., the radius of the Earth? The average radius of the Earth is around 6368 km. This length represents approximately 0.004 % of 1 au and can be neglected without significant loss of accuracy.

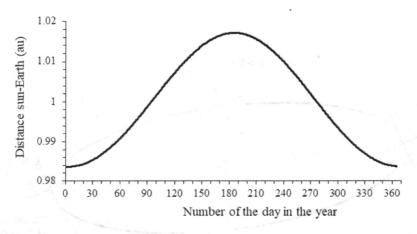

Figure 1.3 Distance between the sun and the Earth as a function of the number of the day in the year.

2 Greif J., Scharmer K. Eds, 2000. *European Solar Radiation Atlas*. Vol. 2: Database and Exploitation Software. Published on behalf of the European Commission by Les Presses de l'Ecole, Ecole des Mines de Paris, France, 296 p. [This book offers a set of equations to compute the position of the sun relative to an observer at ground. This set is reproduced here. C library is available at http://www.oie. mines-paristech.fr/Valorisation/Outils/Solar-Geometry/]

1.3 Solar declination

Note in Figure 1.2 that the equatorial plane of the Earth is tilted relative to the plane of the ecliptic, with a constant angle throughout the revolution. In other words, the axis of rotation of the Earth passing through the two poles is inclined with respect to the normal (perpendicular) to the plane of the ecliptic, with this same angle of inclination. This angle is called the obliquity of the ecliptic. It is about 0.409 rad, or 23.45°, or 23° 26′.

Note in Figure 1.2 that the distances between the sun and the poles vary. At times, the two poles are equidistant from the sun, while at other times, one of the poles is further from the sun than the other. The two points of the extreme orbit where the difference in distances between the sun and each of the poles is greatest are illustrated in Figure 1.4. The diagram on the left represents the case where the distance between the sun and the North Pole is smaller than that between the sun and the South Pole, the sun being located on the right in this diagram. Said differently, this case is the point of the orbit where the north of the axis of the Earth is most tilted toward the sun. This point is called the June solstice. This extreme takes place between 20 and 22 June, depending on the year. In the diagram on the right, the solar rays come from the left. This time, it is the distance between the sun and the South Pole which is shorter than the distance between the sun and the North Pole. Said differently, it is the point of the orbit where the south of the axis of the Earth is most inclined toward the sun. This point is called the December solstice, and this extreme takes place between 20 and 22 December, depending on the year. By extension, the solstices represent also the moments when these points are reached.

Given this difference in distances between the sun and each of the poles, what is its influence on the angle of incidence of the solar rays on a horizontal plane located at the top of the atmosphere? It can be very large depending on the day of the year. Take the case of a point of latitude +45°, represented by a star in Figure 1.4. The angle of incidence is not the same between the left and right diagrams. It is 21.55° (= 45°−23.45°) for the diagram on the left and 68.45° (= 45° + 23.45°) on the right.

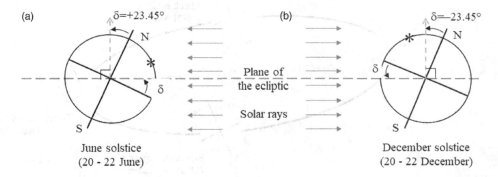

Figure 1.4 Diagram showing the cases where the difference in distance between the sun and each pole is the greatest. (a) The North Pole is the closest to the sun, while it is the South Pole in (b). The Earth is drawn in cross section with the equatorial plane and its rotation axis. The angle δ is the solar declination. The star indicates the point of latitude 45° on the sphere.

This variation in the difference in distances between the sun and the poles during the year can be described by an angle, called solar declination and noted δ. The solar declination is the angle formed, on one side, by the straight line connecting the Earth to the sun and, on the other, by its projection on the equatorial plane, as shown in Figure 1.4. It is positive when the sun is pointing north of the equator (left diagram) and negative when the sun is pointing south of the equator (right diagram). Equivalently, the solar declination is the angle formed by the normal to the plane of the ecliptic and the axis of rotation of the Earth (Figure 1.4).

During an orbit, the solar declination and therefore the angle of incidence of the solar rays arriving at the top of the atmosphere on a horizontal plane vary continuously between the two extremes reached at the solstices (Figure 1.5). The equinoxes are the points of the orbit where δ is zero, in March and September. By extension, the equinoxes represent also the moments when these points are reached. The March equinox takes place between 19 and 21 March, depending on the year, and the September equinox takes place between 21 and 24 September. The declination is positive between the equinoxes of March and September and negative during the other half of the year.

Because δ varies continuously, δ should be calculated at each time. In practice, a daily value provides sufficient precision for δ in many applications. The error made is less than 0.001 rad in absolute value, i.e., 0.06°, or 4′ of arc. Let a be the year. Note num_0 the time between 1 January at 00:00 and the March equinox of this year for longitude 0°. Let ω_{day} be the angle formed by the sun–Earth line for a given day, and that for the day of the March equinox. These quantities are given by the following equations, where the operator INT designates the integer part of a mathematical expression:

$$num_0 = 79.3946 + 0.2422(a - 1957) - \text{INT}\left[(a - 1957)/4\right] \tag{1.7}$$

$$\omega_{day} = (2\pi/365.2422)(num_{day} - num_0) \tag{1.8}$$

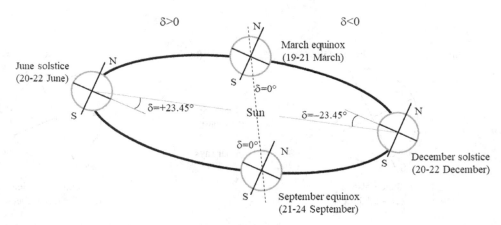

Figure 1.5 Schematic view of the change in solar declination δ during a revolution of the Earth around the sun. The Earth is drawn in cross section with the equatorial plane and its rotation axis. δ is given for both solstices and equinoxes with the corresponding dates. The dashed line divides the part of the orbit where the declination is positive from that where it is negative.

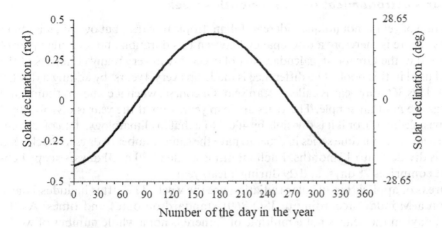

Figure 1.6 Solar declination (in radian and degree) during the year 2019.

δ is calculated as follows:

$$\delta = b_1 + b_2 \sin\left(\omega_{day}\right) + b_3 \sin\left(2\omega_{day}\right) + b_4 \sin\left(3\omega_{day}\right)$$
$$+ b_5 \cos\left(\omega_{day}\right) + b_6 \cos\left(2\omega_{day}\right) + b_7 \cos\left(3\omega_{day}\right) \qquad (1.9)$$

where $b_1 = 0.0064979$; $b_2 = 0.4059059$; $b_3 = 0.0020054$; $b_4 = -0.0029880$; $b_5 = -0.0132296$; $b_6 = 0.0063809$; and $b_7 = 0.0003508$.

Figure 1.6 exhibits the variation of the solar declination according to the number of the day in the year 2019. The declination is zero for days #80 (20 March, March equinox) and #266 (23 September, September equinox). The minimum is reached at the December solstice (day #356, 22 December) and is equal to about -0.409 rad, or $-23.45°$, or $-23° 26'$. The maximum is approximately $+0.409$ rad, or $+23.45°$, or $23° 26'$, and is reached at the June solstice (day #172, 21 June). It is equal to the obliquity of the ecliptic. There are few differences in the declination between the different years.

The extreme values of the declination determine four particular latitudes: the tropics and the polar circles. The latitudes of the Tropic of Cancer and Tropic of Capricorn are respectively the maximum and minimum of the solar declination. At these latitudes, the sun reaches the zenith of the observer on the ground, i.e., at its vertical exactly, at the June solstice at the Tropic of Cancer and at the December solstice at the Tropic of Capricorn. The latitudes of the polar circles are the complement to $\pi/2$ of the obliquity of the ecliptic and are equal to 1.1615 rad ($+66.55°$) in the northern hemisphere and -1.1615 rad ($-66.55°$) in the southern hemisphere. Polar circles are the latitudes at which the sun is above the horizon during 24 h during the day of solstice of June and December, respectively. This will be detailed in the next chapter.

1.4 Definitions of time: from year to second

Time is closely related to the solar cycle and to the two rotations. The orbit of the Earth around the sun and the solar declination define the year and the astronomical seasons and then the months, while the rotation of the Earth on itself with respect to the sun defines the day and then the hour, the minute, and the second.

1.4.1 Year – astronomical season – month – week

The definition of year is not unique: sidereal, Julian, tropical year…; but overall, it is about 365.25 days. There is therefore a discrepancy between this duration and a regular year of 365 days, used in the Gregorian calendar, that of every day in very many countries of the world, and used in this book. The difference is made up every 4 years, by adding a day (29 February). This 366-day year is called a leap year. Of course, when it comes to timing and time, things are not that simple. The years are leap years only if the year is divisible by 4 and not divisible by 100, or if it is divisible by 400. Note that in climatology, data of 29 February are excluded from time series in order to have the same number of days for each year.

A year is divided into 12 months. Each month is made of 30 or 31 days, except February that comprises 28 days, and 29 during a leap year.

Some areas of application of solar radiation use the week. The latter includes 7 days. It starts on a Monday, according to ISO 8601 standard on dates and times. As the number of days in the year is not a multiple of 7, there is not a whole number of weeks in a year: There are 52 or 53 weeks. The ISO standard says that the first week of the year is the week comprising the first Thursday in January, or equivalent, 4 January. Thus, the first Thursday of 2020 being 2 January, week #1 of 2020 started on Monday, 30 December 2019. There were therefore 52 weeks in 2019. What about the week #1 of 2021? The first Thursday of 2021 being 7 January, week #1 of 2021 began on Monday, 4 January 2021. The year 2021 is made of 53 weeks.

The equinoxes and solstices define the beginnings and ends of the astronomical seasons. The March equinox marks the beginning of the astronomical spring. It corresponds to the boreal spring in the northern hemisphere and to the austral (or southern) autumn in the southern hemisphere (Table 1.1). This season ends with the June solstice, which marks the start of the astronomical summer, or boreal summer and austral winter. It is during this season that the Earth is furthest from the sun, with the northern hemisphere closer to the sun than the southern hemisphere. Astronomical summer ends at the September equinox, which marks the beginning of the astronomical autumn, that is, boreal autumn and austral spring, the season ending at the solstice of December. Then begins the astronomical winter, or boreal winter and austral summer. During this season, the Earth is closest to the sun, with the southern hemisphere closer to the sun than the northern hemisphere.

Be careful with the use of the terms *summer* and *winter*. If there is no ambiguity in astronomy, there may be in other areas, depending on the hemisphere. Take the summer solstice. According to some authors, it can denote the solstice marking the start of the astronomical summer. To others, it can designate the solstice marking the start

Table 1.1 Correspondence between astronomical, boreal, and austral seasons

Astronomical season	Period (approximately)	Season in the northern hemisphere	Season in the southern hemisphere
Spring	20 March–21 June	Boreal spring	Austral autumn
Summer	21 June–23 September	Boreal summer	Austral winter
Autumn	23 September–21 December	Boreal autumn	Austral spring
Winter	21 December–20 March	Boreal winter	Austral summer

of summer in the hemisphere concerned, that is, June in the northern hemisphere and December in the southern hemisphere.

The astronomical seasons do not count the same number of days. The astronomical spring lasts about 93 days, the summer about 94 days, the autumn about 90 days, and the winter about 89 days.

Astronomical seasons are defined by variations in solar declination and do not correspond to weather seasons. These are defined by several meteorological variables, such as surface air temperature, rainfall, relative humidity, and sunshine, as well as their variation during the year. If in Europe the seasons are quite similar to the astronomical seasons, it is not the same everywhere in the world. For example, in the equatorial and tropical regions, the seasons are rather divided into dry seasons and rainy, or wet, seasons. At the poles, the year is punctuated by alternating polar days and nights.

Climatologists are often interested in global scales, that is to say, the entire planet or large regions or continents, within which the weather seasons may vary. They have defined four seasons made of full months:

- winter (December, January, February), noted DJF;
- spring (March, April, May), noted MAM;
- summer (June, July, August), noted JJA;
- autumn (September, October, November), noted SON.

Please allow me a digression about the notation of the dates. There are several notations found in the scientific publications or others. For example, 1 August 2021 may also been written as August 1, 2021. If one uses a number for the month, from 1 (January) to 12 (December), 1 August 2021 may be noted 1/8/2021, or 8/1/2021. This can be very confusing as the latter may be read as 8 January 2021. In order to avoid such confusion, I introduce here the ISO standard on the representation of dates and times.[3] The ISO format is applicable if the dates are given in the Gregorian calendar and the hours in a 24-h reference system and not twice 12 h.

The encoding of a date is *YYYY-MM-DD*. The year *YYYY* is coded with four characters, e.g., 2021. The month *MM* is coded with two characters, from 01 (January) to 12 (December). The day *DD* is the number of the day in the month *MM*, from 00 up to 31, and is coded with two characters. For example, 2021-08-01 stands for 1 August 2021. Reduced representations are possible. For example, 2021 stands for the year 2021 without referring to a specific day in this year. 2021-03 represents March 2021 without referring to a specific day in this month. Similarly, 11-12 stands for 12 November without referring to a specific year.

The notation *YYYY-DDD*, where *DDD* is coded with three characters, from 001 up to 366, represents the date under the form of a year and the number of the day in the year. For example, 2021-124 corresponds to 4 May 2021.

---

3 ISO 8601:2004, 2014. *Data Elements and Interchange Formats – Information Interchange – Representation of Dates and Times*, Third edition, 40 p, International Standard Organization, Geneva, Switzerland. [Several programming languages have integrated this standard, such as *datetime* in the Python language.]

A week is encoded as *YYYY-Www*, where *ww* is the number of the week in the year, coded with two characters. A specific day may be given, knowing that Monday is coded 1. For example, 2017-W03 is for the third week of 2017, with referring to a specific day in the week. 2017-W03-3 stands for Wednesday of the third week, that is, Wednesday, 18 January 2017.

There is no approved unit in the international system for the day, week, month, year, or decade (10 days or 10 years). It is therefore better to write in full day, week, decade, month, and year. Note that some authors use the Latin word *annum* for the year with the unit symbol *a*. Others use *yr* for the year.

1.4.2 Day – hour – minute – second

While orbiting around the sun, the Earth rotates on its axis as shown in Figure 1.2. This rotation determines the day, then the hour, the minute, and the second. The direction of rotation is the same than the direction of the orbit, i.e., counterclockwise. The duration of the rotation induces the concept of the mean solar day, which includes 24 h. More exactly, the Earth rotates on its axis in 23 h 56 min and 4 s. During a rotation, the Earth has moved along its orbit. It takes on average 3 min and 56 s to find the sun at the same point on the sky for a total of 24 h.

It takes 60 s to make a minute whose unit symbol is *min*. It takes 60 min to make 1 hour whose unit symbol is *h*. Finally, the mean solar day includes 24 h, i.e., 86,400 s. The second has long been defined as the fraction 1/86,400 of the mean solar day. However, the speed of rotation of the Earth tends to slow down, causing the mean solar day to lengthen. This is discussed later. A new reference for the second has been established for several decades and is based on the count of the number of oscillations of cesium atoms.

There is possible confusion about the term day. Indeed, it can mean either duration of 24 h or the time during which the sun is above the horizon, that is to say, the time interval between sunrises and sunsets. In this book, generally speaking, I use day for duration of 24 h and daytime or daylength for the period of time during which the sun is above the horizon. If there is a risk of confusion, I would be explicit. However, the night does mean the time during which the sun is below the horizon.

According to the ISO 8601 standard on dates and times, the encoding of a time is *Thh:mm:ss*. The letter *T* means that it is the encoding of an hour, and it can be omitted if there is no ambiguity. The hour *hh* varies between 00 and 24 and is coded in two characters. Minutes *mm* and seconds *ss* are coded in two characters from 00 to 60. For example, T13:34:21 stands for 13 h 34 min and 21 s and is equivalent to 13:34:21. Reduced forms are possible: T13:34 or 13:34 means 13 h 34 min without referring to a specific second in this minute. The letter *T* is useful for encoding a date and a time as it separates the date and the time. For example, 2017-05-02T13:34:21 stands for 2 May 2017 at 13 h 34 min 21 s.

Why a base 60 and not decimal for time? The choice of the base 60 was made several millennia ago, and this base is still in use for angles and hours. This does not facilitate the task in calculations, for which it is preferable to express the time in decimal system rather than sexagesimal. For example, the time 10 h 30 min 30 s is easier to handle digitally in the form 10.55 h. Converting a time given in the sexagesimal notation to a decimal one is done by first dividing the number of seconds by 60. This result is added to the number of minutes; the whole is then divided by 60. The result is added to the number of hours, to form the decimal part. For example:

$$14\,h\,34\,min\,12\,s \to 14\,h\,34\,min + 12/60 \to 14\,h\,34.2\,min \to 14\,h + 34.2/60 \to 14.57\,h$$

Conversely, converting a time given in decimal notation to a sexagesimal one is done by first multiplying the decimal part by 60. It yields the number of minutes in decimal format. The decimal part is then multiplied by 60 and yields the number of seconds. For example:

$$14.57\,h \to 14\,h + 0.57 \times 60 \to 14\,h\,34.2\,min \to 14\,h\,34\,min + 0.2 \times 60 \to 14\,h\,34\,min\,12\,s$$

Allow me a brief reminder on units. In science and engineering, the unit symbols are not written in plural, and the numbers and the unit symbols are separated by a space. So, two mistakes are made by writing, for example, *every 2mins*, instead of *every 2 min*.

1.5 Mean solar time – true solar time

The mean solar time (MST) is related to the mean solar day. It is defined in such a way that it is equal to 12:00 on average over the year when the sun reaches its highest during its course in the sky. It is noted t_{MST}. Since the sun cannot be at its highest point simultaneously for all the geographic points of the Earth, it follows that the mean solar time is not the same everywhere and depends on the longitude λ.

Given the mean solar time $t_{MST}(\lambda = 0)$ at longitude 0°, what is the mean solar time $t_{MST}(\lambda)$ at any longitude λ? Remember that the Earth rotates of 360° (2π) on its axis in 24 h, i.e., that the 360° in longitude is covered in 24 h. Consequently, a rotation of 1° is covered in 4 min (= 24 h/360°). Thus, the difference in mean solar time between the longitudes 0° and λ (expressed in degree) is given in h by:

$$t_{MST}(\lambda) - t_{MST}(\lambda = 0) = (24\lambda/360) \tag{1.10}$$

The difference is independent of latitude. A difference of 1 h corresponds to a difference in longitude of 15° (= 1 h/4 (min/°)). For example, it is 12:00 MST in the city of Nice, France (latitude: 43.70°; longitude: 7.27°), about 12 min before the city of Brussels (latitude: 50.85°; longitude: 4.35°). In another example, it is 12:00 MST in the city of Bangkok, Thailand (latitude: 13.75°; longitude: 100.49°), about 20 h before the city of Mexico City, Mexico (latitude: 19.43°; longitude: −99.13°).

If the speed of rotation of the Earth on itself is almost constant, in contrast, the angular speed of the Earth varies slightly during a revolution since the orbit of the Earth around the sun is an ellipse. As a result, the sun does not reach its highest point above the same site at 12:00 MST every day. The peak time may differ by approximately ±15 min around 12:00 MST. As has been done for mean solar time, a true solar time (TST) is defined, noted t_{TST}, so that it is equal to 12:00 when the sun reaches its highest point each day. The true solar time is also known as apparent solar time, or local apparent time.

Take the example of the sundial to illustrate the difference between mean solar time and true solar time. Suppose that I am located at a point of medium latitude and that

I plant obliquely a rigid rod, oriented toward the south in the northern hemisphere or toward the north in the southern hemisphere. Suppose that I make a line every day where the shadow of the stem is the shortest, i.e., when the sun is the highest. This line would correspond to 12:00 TST. The next day, I will find that the new line does not overlap the previous one, and so on. At the end of a year, I would realize that all the positions of the lines are distributed rather symmetrically around an average position that corresponds to 12:00 MST. The sundials have time indications representing the mean solar time, while the shadow of the rod indicates the true solar time.

The difference between t_{TST} and t_{MST} is called equation of time though it is not an equation in the usual sense. As before, let note num_{day} the number of the day in the year and $angle_{day}$ the day angle formed by the sun–Earth line for the day num_{day}, and that for 1 January given in equation (1.4). The equation of time can be expressed in h, with the following approximate formula:

$$t_{TST} - t_{MST} = -0.128\sin\left(angle_{day} - 0.04887\right) - 0.165\sin\left(2\,angle_{day} + 0.34383\right) \quad (1.11)$$

A positive difference means that the sun is ahead of the mean solar time; i.e., the sun is further west than the mean solar time indicates. On the contrary, a negative equation of time indicates that the sun is behind the mean solar time, that is to say, more to the east than the latter indicates.

Figure 1.7 depicts the equation of time as a function of the day in the year. The shape of the equation of time is approximately the composition of two sinusoids. The first reflects the influence of the eccentricity of the orbit; it has a period of 1 year and cancels at the perihelion, approximately between 2 and 5 January, as well as at the aphelion, approximately between 3 and 5 July. The second reflects the obliquity of the ecliptic; it has a period of 0.5 year and cancels at the equinoxes and solstices.

The maximum of the equation of time is reached around 31 October and is equal to 0.276 h, or 16.57 min, or 16 min and 34 s. The minimum is equal to −0.242 h and is reached around 13 February, or −14.52 min, or −14 min 31 s. The equation of time is 0

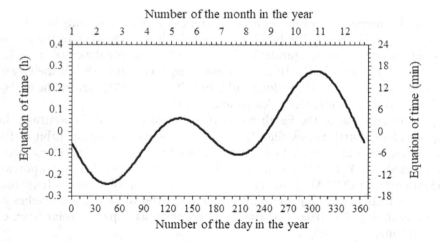

Figure 1.7 Equation of time (t_{TST}–t_{MST}) in h (left axis) and min (right axis) as a function of the day in the year.

on approximately 16 April, 14 June, 31 August, and 25 December. There is a secondary maximum of 0.061 h (3.66 min) around 14 May and a secondary minimum of 0.106 h (6.36 min) around 27 July.

1.6 Coordinated universal time and legal time

1.6.1 Coordinated universal time (UTC)

Human activities need a common time frame on an international scale. International atomic time is determined by a network of a few hundred atomic clocks, the same as those used to define the second. This time is extremely stable. However, it does not exactly follow the rhythm of the rotation of the Earth on itself, which is the basis of legal time.

The mean solar time at longitude 0° has been the international reference for time for long. This longitude is also called the Greenwich meridian, after the name of the British observatory located at this meridian. The reference was called the GMT (Greenwich mean time). Since the 1970s, in order to improve accuracy, the reference time has been determined by observing stars outside the solar system, under the responsibility of the International Earth Rotation and Reference Systems Service, abbreviated as IERS. This determination by star sighting is called universal time UT1, and its precision is 4 ms.

The speed of rotation of the Earth on itself is not constant. High-frequency irregularities in speed occur due to the displacement of air and water masses, or even to earthquakes. They are unpredictable, and this makes impossible a very accurate definition of time based solely on this rotation, and therefore on UT1. The international atomic time is used to smooth the variations due to irregularities in rotation, as it sets the frequencies of time, like the second. This yields the coordinated universal time, curiously abbreviated as UTC.

As the speed of rotation of the Earth decreases slowly in particular under the effects of the tides and since time UT1 describes this rotation, UT1 increases in time. Consequently, the difference between UT1 and international atomic time grows over time. To avoid too great a disparity between the UT1 time and the UTC, the IERS proceeds to an adjustment of the UTC by inserting an entire second, called leap second, so that the UTC does not deviate no more than 0.9 s from UT1 time. There is therefore sometimes a minute including 61 s in these catch-up operations, which take place on 30 June or 31 December. At the time of writing, the last adjustment took place on 2016-12-31 (31 December 2016) at 23:59:60 UTC; 27 leap seconds have been added since the start of the corrective system in 1972. In general, the date of the next adjustment cannot be known with exactitude.

Let t_{UTC} be a moment in UTC at a geographic location of longitude λ. The mean solar time t_{MST} is given in h by:

$$t_{MST} = t_{UTC} + (24\lambda/2\pi) \quad \text{if } \lambda \text{ is in rad}$$

$$t_{MST} = t_{UTC} + (24\lambda/360) \quad \text{if } \lambda \text{ is in degrees}$$

For example, the mean solar time is equal to the coordinated universal time at longitude 0°. It is equal to the coordinated universal time plus 12 h at longitude 180° and equal to the coordinated universal time minus 8 h at longitude −120°.

1.6.2 Legal time – civil time – time zone

Legal time t_{legal} is the time legally used in a country. It is also called local time, standard time, local standard time, or civil time. It is based on the mean solar time using time zones. The longitudes of the Earth are subdivided into 24 sets with a width of 15° each (= 360°/24 h). Each set is assigned the mean solar time of central longitude. This is the principle of time zones. Each time zone represents an increase or decrease of 1 h in mean solar time compared to its neighbor. The zone centered on the longitude 0° has UTC. The longitude 180° corresponds to the UTC + 12 h time zone, known as the date change.

The legal time at a given place is often the time of the time zone containing it. For practical reasons, the limits of a time zone often follow the borders of all the countries contained therein. If the size of a country is such that it covers more than 15° in longitude, then the country can adopt several legal hours. In general, countries tend to use a time zone in such a way that the mean solar time in their territory is close to the legal time (i.e., for example, that the solar noon is not too far from the legal noon). States set the legal time on their territory using a fixed offset from the coordinated universal time. This offset is most often equal to an integer number of hours (time zone principle), but several states have chosen a half-hour offset, for example, the central part of mainland Australia (+9.5 h), Afghanistan (+4.5 h), India (+5.5 h), Iran (+3.5 h), or Myanmar (+6.5 h), or even a quarter of an hour, for example, Nepal (+5.75 h) or the Eucla region in mainland Australia (+8.75 h). The date change line is not exactly the longitude 180° and follows the borders of states to avoid the day change within the same state.

Very large countries are often divided into several time zones in order to avoid too great a difference between the legal time and the mean solar time. For example, Russia covers approximately longitudes from 30° to 190° and includes eight time zones. This is not a rule observed by all. For example, China extends approximately from 75° to 135° in longitude, and the mean solar time at extreme longitudes is respectively t_{UTC} + 5 h and t_{UTC} + 9 h. Four zones could have been adopted. The same legal time prevails throughout China: t_{UTC} + 8 h, i.e., approximately the mean solar time of the capital Beijing, whose longitude is 116.36°. The difference between the legal time and the mean solar time can be very large depending on the city. Suppose a legal time in China equal to 12:30. At the longitude of Beijing, the corresponding UTC is then $t_{UTC} = (12.5 - 8)$ h, i.e., 04:30, and the mean solar time is 12:16. The city of Chengdu, of longitude 104.06°, is located more to the west than Beijing; although the legal time is the same, 12:30, the mean solar time in Chengdu is 11:27. It equals 10:21 in the city of Urumqi even further west whose longitude is 87.62°.

West African states such as Mauritania and Senegal could have adopted t_{UTC} + 1 h as legal time given their longitudes, but preferred t_{UTC}. A contrary example is given by the countries of the Iberian Peninsula (Spain and Portugal) which have the same longitudes as Ireland and the United Kingdom. While Portugal, Ireland, and the United Kingdom have t_{UTC} as legal time, Spain has adopted t_{UTC} + 1 h. In the Pacific Ocean, some states, made up of very distant islands, covered several time zones and notably overlapped the date change line (t_{UTC} + 12 h). It is difficult to administer a state, one part of which does not have the same date as the other. These states have adopted legal times, ranging from t_{UTC} + 13 h to t_{UTC} + 14 h, to remedy this difficulty. As a result, these states have a 24 h offset from the areas using $t_{UTC}-11$ h and $t_{UTC}-10$ h. If it is Friday noon in the former, it is Thursday noon in the latter.

Several countries have adopted summer and winter time systems, which are often created to maximize the use of natural light. This system is called daylight saving time. It consists in increasing the time difference with the UTC by adding 1 h to the legal time during a period going approximately from the equinox of March to that of September in the northern hemisphere and, conversely, from the equinox of September to that of March in the southern hemisphere. In some federal countries, such as Australia, states may adopt this system and others may not. To add an extra dose of complexity, the dates of change from summer time to winter time or vice versa are not the same in the world.

Reader, you got it. Legal time at a particular place is not always easy to understand, especially since this time is governed by local political considerations that may evolve under different influences. For example, Georgia changed from $t_{UTC} + 4$h to $t_{UTC} + 3$h on 27 June 2004, and then back to $t_{UTC} + 4$h on 27 March 2005. Another example is given by the Phoenix Islands and the Line of Islands, part of Kiribati in the Pacific Ocean, which skipped 31 December 1994, jumping directly from 30 December to 1 January, and respectively from $t_{UTC} - 11$h to $t_{UTC} + 13$h and from $t_{UTC} - 10$h to $t_{UTC} + 14$h. Samoa and Tokelau, in the Pacific Ocean, did the same, skipping 20 December 2011, and going from $t_{UTC} - 11$h to $t_{UTC} + 13$h. The last example is Crimea, part of Ukraine, which went from $t_{UTC} + 2$h to $t_{UTC} + 3$h on 18 March 2014, after its annexation by Russia.

Why do I insist so much on legal time? The legal time is very often that used for the time-stamping of solar radiation measurements. As the calculations of the solar radiation must be made with the true solar time, it is important to know well the relation between the legal time and the UTC, and further the mean solar time, then the true solar time using the equation of time.

Fortunately, a lot of information on the legal time adopted by each country is available on the Web. You should not hesitate to do extensive research to be sure of the meaning of time stamps, or even contact the operator of the measuring station whose measurements you want to use. In addition, it is quite common that the legal time used for time-stamping does not take into account the time change between winter and summer, which makes things easier.

The ISO standard makes it possible to represent different time systems. Without mention, it is local time. The letter Z is added to indicate the UTC. For example, T13:34:21Z means 13 h 34 min 21 s UTC. The use of legal time is possible by indicating the difference between this time and UTC. The representation of this difference, expressed in *hh:mm*, is added to that of the hour in local time. For example, T13:34:21+01:00 means 13 h 34 min 21 s in legal time which is ahead of UTC by 1 h. In another example, T13:34:21−01:45 means 13 h 34 min 21 s in local time, which is behind UTC by 1 h 45 min.

I mention in passing the existence of a time zone system used by the armed forces of different countries. This system is equivalent to that of time zones, and their names use the phonetic alphabet of the North Atlantic Treaty Organization (NATO). The time called Z (as zero meridian and coded as Zulu) corresponds to UTC. The 12 time zones east of "Zulu" receive the letters from A ($t_{UTC} + 1$h) to M ($t_{UTC} + 12$h), and those to the west of "Zulu", the letters from N ($t_{UTC} - 1$h) to Y ($t_{UTC} - 12$h). The letter J is reserved and indicates the local time of the observer.

1.6.3 To sum up: converting the legal time to the true solar time

In summary, if one wishes to calculate the position of the sun in a given place whose geographic coordinates are known, at a given instant, one must know the true solar time for this instant. If the time is given in legal time, then the following operations must be carried out to obtain the true solar time:

- convert the legal time to UTC, taking into account the time difference adopted by the country and possibly summer and winter hours;
- then, use equation (1.12) to calculate the mean solar time; and
- finally, use equation (1.11) to obtain the true solar time.

Chapter 2

The course of the sun over an observer on the ground

At a glance

The sun is so far from the Earth that it appears as a sphere of small diameter; its apparent diameter is about $32' \pm 0.5'$ of arc, or about half a degree. Therefore, the rays arriving at the top of the atmosphere can be considered to be parallel.

The solar zenithal angle is the angle formed by the direction of the sun and the local vertical. The azimuth of the sun, or solar azimuthal angle, is the angle formed by the projection of the direction of the sun on the horizontal plane and the north. The azimuth increases clockwise.

The direction of the sun perceived at the top of the atmosphere is very close to that perceived by an observer on the ground; the refraction of solar rays by the atmosphere induces a very slight variation that can be neglected. Consequently, the solar angles are nearly identical whether on the ground or at the top of the atmosphere, or more generally, at any altitude.

The rotation of the Earth on its axis determines the duration of day and night. To avoid confusion with the fact that the day actually lasts 24 h, daytime, or day-length, is defined here as the period during which the sun is above the horizon, i.e., between sunrise and sunset. Night is the period when the sun is below the horizon. These definitions are taken in the astronomical sense, that is to say, without obstacle in the line of sight. Otherwise, as, for example, in a city or in an area with a relief, the effective daytime may be different from the astronomical one.

The notion of effective solar angles is a means of associating unique angles with each measurement that has been taken over a certain time interval.

At any time and in any place, the solar radiation received on the ground depends on two main elements: the terrestrial atmosphere and the radiation from the sun arriving at the top of the atmosphere, also called extraterrestrial radiation. The atmosphere depletes the solar radiation as the latter makes its way downward to the ground. Extraterrestrial radiation is therefore the driving quantity because radiation at ground level cannot exceed it, except in rare conditions. The extraterrestrial radiation received by a surface at any time and in any place is a function of the day in the year, that is to say, of the position of the Earth on its orbit and of the solar declination, of the time in the day, of the latitude, and of the inclination of the surface relative to the horizontal. The direction of the sun determines the angle of incidence of the solar rays onto a surface at the top of the atmosphere. The previous chapter has shown the importance of this angle in the transfer of energy by radiation between the incident radiative energy and the receiving plane. The greater the angle of incidence, the lower the energy received by the plane.

This chapter deals with the calculation of the direction of the sun and more generally with the course of the sun over an observer at ground. The relative geometry between the direction of the sun and an observer on the ground, in other words the apparent course of the sun in the sky above an observer, determines the angle of incidence. Therefore, the direction of the sun as seen by the observer must be known accurately. It is described by the solar zenithal angle, or equivalently the angular elevation above the horizon, and the solar azimuth. These angles are defined in this chapter.

Several examples are given. A series of simple but fairly accurate equations is provided to calculate the various elements discussed. They are those of the European Solar Radiation Atlas (ESRA). The calculation errors of the solar angles are less than 0.2° for the solar zenithal angle and 0.4° for the azimuth. The accuracy in time is of the order of a few minutes, and these equations are suitable if one is interested in radiation values on timescales of hour or larger. For better accuracy, more accurate equations should be preferred.[1]

The mathematical expression of the course of the sun in the sky makes it possible to calculate the solar angles at sunrise and sunset. From these results, hours of sunrise and sunset can be calculated, and therefore the length of the day, called daytime or daylength, and that of night.

Since the radiation measurements are made over a certain period of time, during which the solar angles change, what are the solar angles to be assigned to each measurement? This chapter answers this problem by presenting the concept of effective solar angles which is a means of associating unique angles with each measurement that has been taken over a time interval greater than 1 min, while the solar angles are only defined for very short moments, less than 1 min. Solutions for associating the solar angles with each measurement are discussed.

Given the enormous size of the sun, is there a single direction of the sun seen from Earth or more than one? In fact, the sun is so far from the Earth that despite its size, it appears as a very small sphere to an observer on the ground or at the top of the atmosphere. For many practical purposes, the sun can be thought of as a point, and there is only one direction of the sun. One consequence is that the solar rays arriving at the top of the atmosphere can be considered as parallel. Mathematically, it is a little different because the solid angle under which the sun appears is not zero and is equal to $0.68 \ 10^{-4}$ sr. In other words, the apparent diameter of the sun is about $32' \pm 0.5'$ of arc, or about half a degree ($0.53° \pm 0.008°$ of arc). The solar rays arriving at the same point at the top of the atmosphere from all points included in this solid angle are not

---

1 Software libraries are available in C and Matlab at rredc.nrel.gov/solar/codesandalgorithms/spa (SPA algorithm) and http://www.oie.mines-paristech.fr/Valorisation/Outils/Solar-Geometry/ (SG2 algorithm), and also in Python at pysolar.org (PySolar) or https://rhodesmill.org/skyfield/ (SkyField).

perfectly parallel. Nevertheless, assuming that the direction of the sun is that of its center and that the solar rays are parallel, this results in very small errors on the solar radiation which are perfectly acceptable in practice.

2.1 Position of the sun seen by an observer on the ground

An observer on the ground perceives a movement different from that described above in Figure 1.2 (Chapter 1). He does not feel the movement of the earth around the sun. For him, the sun describes an apparent movement in the sky, also called the course of the sun. For most latitudes, the sun rises approximately in the east, ascends into the sky until solar noon, then descends, and sets approximately in the west. The exact course depends on the latitude, the time of day, the day in the year and, somewhat, the year. Figure 2.1 is a schematic view of this course for an observer located in the northern hemisphere. The sun is in the south of the observer at solar noon. If I had represented an observer in the southern hemisphere, I could have used the same diagram by reversing the symbols N and S.

The hour angle ω is a very convenient quantity for calculating the position of the sun in its apparent path. Knowing that the sun is at its highest at solar noon, this angle measures the trajectory of the sun between this highest point and the point corresponding to an instant t (Figure 2.2). Said differently, ω measures the angular arc between the plane formed by the vertical and by the longitude of the location of the observer P, and the position of the sun at time t.

The hour angle ω is zero at solar noon. It is counted negatively in the morning and positively in the afternoon. ω describes the course of the sun during a day of 24 h, including night. Therefore, ω ranges between $-\pi$ (midnight solar time) and $+\pi$ (midnight but 24 h later). In Figure 2.2, the hour angle at the sunrise is noted $\omega_{sunrise}$ (negative) and that at the sunset is noted ω_{sunset} (positive). These angles are calculated in the later section.

The definition of the hour angle is independent of the hemisphere and is the same for the southern hemisphere. In Figure 2.2, the point P is located in the northern hemisphere. For a point in the southern hemisphere, Figure 2.2 would be the same, except that the north should be interchanged with the south.

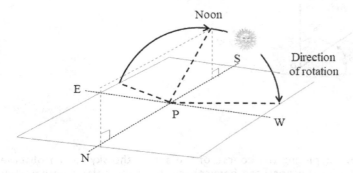

Figure 2.1 Schematic view of the course of the sun in the sky for an observer located at P in the northern hemisphere.

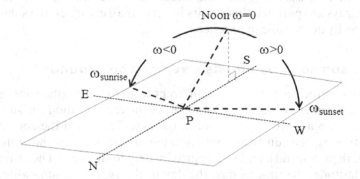

Figure 2.2 Hour angle ω of the sun for an observer located at P in the northern hem-isphere. $\omega_{sunrise}$ is the hour angle of sunrise, and ω_{sunset} is that of sunset ($\omega_{sunset} = -\omega_{sunrise}$).

As a first approximation, the speed of rotation of the Earth on itself is uniform. Since ω goes around 2π during 24 h, the relationships between ω (in rad) and the true solar time t_{TST} (expressed in h) are:

$$\omega = (2\pi/24)(t_{TST} - 12) = (\pi/12)(t_{TST} - 12) \tag{2.1}$$

$$t_{TST} = 12(1 + \omega/\pi) \tag{2.2}$$

2.1.1 Solar zenithal angle – solar elevation – azimuth

The apparent position of the sun can be described by two angles: the solar zenithal angle θ_S and the azimuthal angle ψ_S, or azimuth (Figure 2.3).

The solar zenithal angle θ_S is the angle formed by the direction of the sun and the local vertical (Figure 2.3). It reaches its minimum when the sun is at its highest at

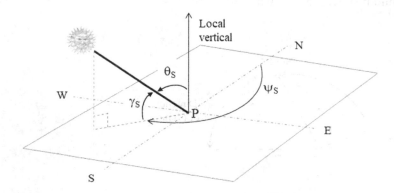

Figure 2.3 Solar angles depicting the course of the sun in the sky for an observer located at P in the northern hemisphere. θ_S is the solar zenithal angle, γ_S is the solar elevation and ψ_S is the azimuth. N, E, S and W identify the cardinal points north, east, south, and west, respectively.

solar noon. If the sun is at the vertical of the place, i.e., if it is at the zenith of the observer, then θ_S is zero. The angle θ_S increases when the sun goes down on the horizon and reaches its maximum at sunrise and sunset.

The definition of this angle corresponds to that of the angle of incidence for a flat horizontal surface at the top of the atmosphere. These two angles are the same in this case.

Assuming that the sun seen by the observer on the ground is a point and not a sphere, and neglecting the refraction of solar rays by the atmosphere, the maximum of θ_S is $\pi/2$ (90°). In fact, on the one hand, the apparent diameter of the sun is about 32′ of arc, or half a degree. The limb of the sun, that is to say, the edge of the solar disk, is therefore 16′ of arc above the horizon when the center of the sun passes under the horizon; solar rays are therefore perceived by the observer. On the other hand, the atmosphere refracts the solar rays, and for this reason, the solar rays do not have a rectilinear trajectory during the crossing of the atmosphere, because of the variation of the density of the air with altitude. The greater the length of the atmosphere crossed, or optical path, the more intense the refraction. The error made by neglecting the refraction is zero when $\theta_S = 0°$, about 1′ of arc when $\theta_S = 45°$, and about 34′ of arc when the sun is on the horizon. By the combination of the two effects (sphere and refraction), the solar rays emitted by the limb of the sun are visible as long as the center of the sun is within 50′ of arc (= 16′ + 34′) below the horizon. The maximum of θ_S is then 1.5853 rad (90.83°). The maximum relative error made on θ_S by neglecting these two effects is of the order of 1 %.

A practical consequence of this calculation is that the relative error in the perception of the direction of the sun by an observer is almost zero and is at most 1 % on the horizon, at times of sunrise and sunset. In other words, two observers, one on the ground and the other at the top of the atmosphere, see the sun in the same direction. Consequently, the solar angles are nearly identical on the ground and at the top of the atmosphere and, more generally, at any altitude.

Another angle is often used instead of the solar zenithal angle: the solar elevation angle, also called the solar altitude angle and noted γ_S. This angle is formed by the direction of the sun and the horizon assuming that there is no obstacle and no relief. It is zero when the sun is on the horizon (sunrise and sunset) and reaches its maximum when the sun is at its highest. The solar elevation is more intuitive than the solar zenithal angle, because it is this elevation that you perceive when you observe the course of the sun. The two angles θ_S and γ_S are complementary to each other:

$$\theta_S + \gamma_S = \pi/2 \tag{2.3}$$

$$\cos(\theta_S) = \sin(\gamma_S) \tag{2.4}$$

$$\sin(\theta_S) = \cos(\gamma_S) \tag{2.5}$$

At a given time t_{TST} in true solar time, the solar zenithal angle θ_S is given by:

$$\cos(\theta_S) = \sin(\Phi)\sin(\delta) + \cos(\Phi)\cos(\delta)\cos(\omega) \tag{2.6}$$

where ω is given by equation (2.1), Φ is the latitude, and δ is the solar declination.

The second solar angle is the azimuth Ψ_S, or solar azimuthal angle. It is defined as the angle formed by the projection of the direction of the sun on the horizontal plane and the north (Figure 2.3). The azimuth ranges from 0 to 2π and increases clockwise. It is 0 northward, $\pi/2$ (90°) eastward, π (180°) southward, and $3\pi/2$ (270°) westward.

The definition of the reference direction for azimuth is arbitrary. Here, it is from the north, as recommended by the ISO 19115 standards for geographic information.

The equations below relate the two solar angles θ_S and Ψ_S:

$$\sin(\Psi_S)\sin(\theta_S) = -\cos(\delta)\sin(\omega) \tag{2.7}$$

$$\cos(\Psi_S)\sin(\theta_S) = \sin(\delta)\cos(\Phi) - \cos(\delta)\sin(\Phi)\cos(\omega) \tag{2.8}$$

If θ_S is different from 0 and if the latitude Φ is different from $\pm\pi/2$, the solar azimuth Ψ_S is given by the following equations:

$$\cos(\Psi') = \left[\sin(\delta)\cos(\Phi) - \cos(\delta)\sin(\Phi)\cos(\omega)\right]/\sin(\theta_S) \tag{2.9}$$

$$\text{If} \quad \sin(\omega) \le 0, \text{then } \Psi_S = \cos^{-1}(\Psi') \tag{2.10}$$

$$\text{If} \quad \sin(\omega) > 0, \text{then } \Psi_S = 2\pi - \cos^{-1}(\Psi')$$

where $\cos^{-1}()$ is the arc cosine function. The azimuth is unknown when $\theta_S = 0$, and it can be set to π by convention. At the poles, Ψ_S is unknown for any position of the sun; any value can be taken. The following equations may have an interest in subsequent calculations involving the azimuth Ψ_S:

$$\sin(\Psi_S) = -\cos(\delta)\sin(\omega)/\sin(\theta_S) \tag{2.11}$$

$$\tan(\Psi_S) = -\sin(\omega)/\left[\sin(\Phi)\cos(\omega) - \cos(\Phi)\tan(\delta)\right] \quad \text{in the northern hemisphere}$$

$$\tan(\Psi_S) = \sin(\omega)/\left[\sin(\Phi)\cos(\omega) - \cos(\Phi)\tan(\delta)\right] \quad \text{in the southern hemisphere}$$

Other standards than the ISO standard are possible for azimuth. For example, several publications in solar engineering have adopted the following standard. Let the azimuth be noted $\Psi_{engineering}$ in this standard. It is measured from south in the northern hemisphere and is positive toward the west. Thus, it behaves similarly to the hour angle ω: It is positive during the afternoon (west) and negative during the morning (east). The solar azimuth $\Psi_{engineering}$ is 0 southward, $\pi/2$ westward, π northward, and $-\pi/2$ eastward. In the southern hemisphere, it is counted from the north and is positive westward. It is 0 northward, $\pi/2$ westward, π southward, and $-\pi/2$ eastward. The azimuth $\Psi_{engineering}$ is equal to $(\Psi_S-\pi)$ in the northern hemisphere and $(-\Psi_S)$ in the southern hemisphere where Ψ_S is defined as above following the ISO 19115 standard.

In summary, given the true solar time and the geographic coordinates of the site of interest, the solar angles are obtained as follows:

- use equation (2.1) to obtain the hour angle from the true solar time,
- then, compute the solar zenithal angle with equation (2.6), and
- finally, compute the azimuth with equation (2.10).

2.1.2 Direction of the sun in the case of an inclined plane

In the previous section, I have computed the solar zenithal angle that gives the direction of the sun with respect to the vertical in the case of a horizontal surface. In practice, surfaces are often inclined: natural slopes, hillsides, photovoltaic panels, roofs, windows... How to compute the angle between the direction of the sun and the normal to an inclined plane? An inclined plane is described with two angles: the inclination β and the azimuth α (Figure 2.4). The inclination β varies between 0 in the case of a horizontal plane and $\pi/2$ in the case of a vertical plane. The azimuth α of the plane, also called orientation, is the angle between the projection of the normal to the plane at the surface of the Earth and the north direction, similarly to the azimuth of the sun.

The angle θ under which the sun is seen is formed by the normal to the plane and the direction of the sun. It is given by:

$$\cos(\theta) = \cos(\beta)\cos(\theta_S) + \sin(\beta)\sin(\theta_S)\cos(\Psi_S - \alpha) \tag{2.12}$$

This and the following equations are valid regardless of the convention adopted for the azimuth. It is important to use the same convention for the azimuth of the plane α and that of the sun Ψ_S. Assume someone has adopted the above-mentioned solar engineering convention: a measurement of the solar azimuth $\Psi_{engineering}$ from the south in the northern hemisphere and counted positively to the west. If this person describes the orientation α of a photovoltaic panel as being equal to 62°, that is to say, that the panel is oriented toward the southwest, the azimuth of the panel in the ISO convention is 242° (= 180° + 62°). If the orientation of the panel is −118°, toward the northeast, then the azimuth of the panel is 62° (= 180°−118°) in the ISO convention. If the panel is located in the southern hemisphere with an orientation of 62°, toward the northwest,

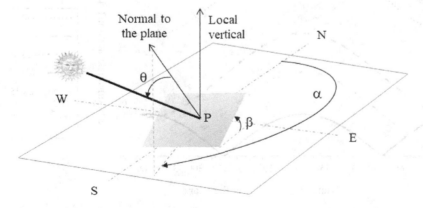

Figure 2.4 Angle θ under which the sun is seen relative to the normal to an inclined plane. The inclination and azimuth of the plane are β and α, respectively.

the orientation is 298° (= 360°−62°) in the ISO convention. If the orientation of this panel is −118°, toward the southeast, the azimuth is 118° in the ISO convention.

It is convenient to express θ as an explicit function of ω for easier calculations for inclined planes. By exploiting (2.6)–(2.12) and given that

$$\cos(\Psi_S - \alpha) = \cos(\Psi_S)\cos(\alpha) + \sin(\Psi_S)\sin(\alpha) \tag{2.13}$$

it comes:

$$\cos(\theta) = A\cos(\omega) + B\sin(\omega) + C \tag{2.14}$$

where

$$A = \cos(\delta)\big[\cos(\Phi)\cos(\beta) - \sin(\Phi)\sin(\beta)\cos(\alpha)\big]$$

$$B = -\cos(\delta)\sin(\beta)\sin(\alpha) \tag{2.15}$$

$$C = \sin(\delta)\big[\sin(\Phi)\cos(\beta) + \cos(\Phi)\sin(\beta)\cos(\alpha)\big]$$

2.2 Solar zenithal angles at solar noon during a year

Figure 2.5 depicts variations in θ_S and γ_S at solar noon during a year at several latitudes. These are the curves of the minimum of the solar zenithal angle and therefore of the maximum of the solar elevation angle, during the year at these latitudes. From these curves, you can quickly make a rough estimate of these angles for other times of the day and at other latitudes.

At the poles, the solar zenithal angle at noon is equal to 66.55° at the summer solstice (June at North Pole and December at South Pole). Then, it increases and reaches 90° and more (sun below the horizon) between the following equinoxes, i.e., between September and March at North Pole and March and September at South Pole.

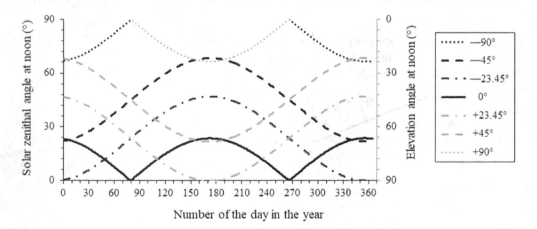

Figure 2.5 Solar zenithal angles and elevation angles at solar noon during a year at latitudes −90°, −45°, −23.45°, 0°, +23.45°, +45°, and +90°.

At latitudes $-45°$ and $45°$, θ_S increases from $21.55°$ ($\gamma_S = 68.45°$) at the summer solstice up to $68.45°$ ($\gamma_S = 21.55°$) at the other solstice, before decreasing.

The sun is at zenith ($\theta_S = 0$, $\gamma_S = 90°$) during the two equinoxes at equator, as well as at all latitudes when $\left|\Phi - \delta(t)\right| = 0$. For example, the sun reaches zenith at December solstice at the Tropic of Capricorn (latitude $-23.45°$). It happens at the Tropic of Cancer at June solstice (latitude $+23.45°$).

The solar zenithal angle reaches its minimum (maximum elevation angle) at solar noon, i.e., when $\omega = 0$. In these conditions,

$$\cos(\theta_S) = \sin(\Phi)\sin(\delta) + \cos(\Phi)\cos(\delta) \tag{2.16}$$

or

$$\cos(\theta_S) = \cos(\Phi - \delta), \quad \text{i.e.} \, \theta_S = \left|\Phi - \delta\right| \tag{2.17}$$

Differently said, the solar zenithal angle at noon is equal to the absolute value of the difference between the latitude and the solar declination.

2.3 Times of sunrise and sunset

Several times offer special interest, such as the times of sunrise and sunset. As seen previously, the hour angles at sunrise and sunset are denoted $\omega_{sunrise}$ (negative) and ω_{sunset} (positive), respectively (Figure 2.2), with $\omega_{sunrise} = -\omega_{sunset}$ if there is no obstacle on the horizon. ω_{sunset} at sunset is computed by setting the solar zenithal angle θ_S to $\pi/2$ in equation (2.6), i.e.,

$$0 = \sin(\Phi)\sin(\delta) + \cos(\Phi)\cos(\delta)\cos(\omega) \tag{2.18}$$

ω_{sunset} is the solution of the equation. It comes:

if $\Phi = \pi/2$, if $\delta > 0$, $\omega_{sunset} = \pi$, otherwise $\omega_{sunset} = 0$

if $\Phi = -\pi/2$, if $\delta > 0$, $\omega_{sunset} = 0$, otherwise $\omega_{sunset} = \pi$

if $\left(-\tan(\Phi)\tan(\delta)\right) \geq 1$, $\omega_{sunset} = 0$ (2.19)

if $\left(-\tan(\Phi)\tan(\delta)\right) \leq -1$, $\omega_{sunset} = \pi$

otherwise $\omega_{sunset} = \cos\left(-\tan(\Phi)\tan(\delta)\right)$

where $\cos^{-1}()$ is the arc cosine function. $\omega_{sunset} = 0$ means that the sun is always below the horizon, while $\omega_{sunset} = \pi$ means that the sun is always above the horizon. The hour angle $\omega_{sunrise}$ at sunrise is given by:

$$\omega_{sunrise} = -\omega_{sunset} \tag{2.20}$$

Azimuths at sunrise and sunset, $\Psi_{sunrise}$ and Ψ_{sunset} respectively, are computed by equation (2.10), where $\theta_S = \pi/2$. $\Psi_{sunrise}$ is less than π, while Ψ_{sunset} is greater than π.

Times at sunrise $t_{sunrise}$ and sunset t_{sunset} are given in true solar time (in h) by equation (2.2):

$$t_{sunrise} = 12\left(1 + \omega_{sunrise}/\pi\right)$$ (2.21)

$$t_{sunset} = 12\left(1 + \omega_{sunset}/\pi\right)$$

By way of illustration, Figure 2.6 gives the time of sunrise in TST, MST, and UTC as a function of the number of the day in the year at longitude 5° and at two latitudes: 0° and 45°. I took the longitude arbitrarily at 5° to properly separate the time in MST and UTC since at longitude 0°, the two times are identical. In this example, the difference between the times MST and UTC is 20 min. At the equator, sunrise time is always 06:00 TST. Times in MST and UTC fluctuate, depending on the equation of time. At latitude 45°, the amplitude of the variation in the sunrise time is greater than at equator. The sun rises between 4.3 h (04:20) and 7.7 h TST (07:42). As in the case of the equator, the time in MST varies around the time in TST depending on the equation of time and similarly for the UTC with an offset of 20 min.

The equator is special latitude: $\omega_{sunrise}$ is always equal to $-\pi/2$ ($-90°$). Sunrise and sunset always take place at the same time in TST, irrespective of the date: $t_{sunrise}$ is 06:00 TST and t_{sunset} is 18:00 TST. At other latitudes, except the poles, $\omega_{sunrise}$ is equal to $-\pi/2$ at the equinoxes, whatever the latitude. During these 2 days, $t_{sunrise}$ is 06:00 TST and t_{sunset} is 18:00 TST, similar to the equator.

At the North Pole, $\omega_{sunrise}$ is equal to 0 during a 6-month period, centered on the December solstice: It is polar night and the sun is always below the horizon. On the contrary, $\omega_{sunrise}$ is equal to $-\pi$ during the 6 other months, centered on the June solstice, and the sun is always above the horizon, and there is no night. It is polar day, sometimes called the midnight sun. The same is happening at the South Pole but with

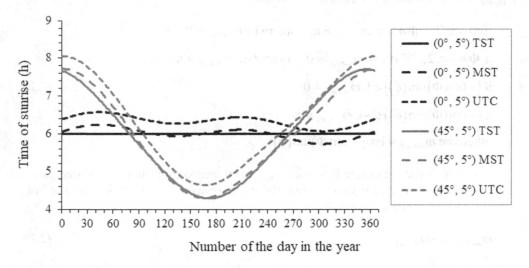

Figure 2.6 Time at sunrise in TST, MST, and UTC during the year, at longitude 5° at two latitudes: equator and 45°.

a 6-month polar night centered on the June solstice and one polar day during the other period centered on the December solstice.

The polar circles are particular latitudes with respect to sunrise and sunset times. The Arctic circle (northern polar circle) is the lowest positive latitude where $\omega_{sunrise} = -\pi$ for at least 1 day, i.e., the lowest positive latitude for which the sun is always above the horizon during 24 h, which occurs during the day of the June solstice. Conversely, during the day of the December solstice, $\omega_{sunrise} = 0$; i.e., the sun is always below the horizon. North of the Arctic circle, as the latitude increases toward the North Pole, there are more and more days of 24 h during which the sun is always below or above the horizon. On the contrary, there is no such day south of the Arctic circle. The situation is similar in the southern hemisphere. The Antarctic circle is the lowest latitude where $\omega_{sunrise} = -\pi$ for at least 1 day, i.e., for which the sun is always above the horizon during 24 h, which occurs during the day of the December solstice. Conversely, $\omega_{sunrise} = 0$ during the day of the June solstice; i.e., the sun is always below the horizon. South of the Antarctic circle, as the latitude increases toward the South Pole, there are more and more days of 24 h during which the sun is always below or above the horizon. On the contrary, there is no such day north of the Antarctic circle.

The calculation of $\omega_{sunrise}$ and ω_{sunset} above does not take into account the effects of the refraction of solar rays by the atmosphere, nor the angular dimension of the solar disk, which make the solar disk disappear from our view only when the center of the disk is below the horizon with an angle of 50′ (0.01454 rad, or 0.8333°). Taking these effects into account, ω_{sunset} is the solution to the following equation:

$$\sin(-0.014544) = \sin(\Phi)\sin(\delta) + \cos(\Phi)\cos(\delta)\cos(\omega_{sunset}) \tag{2.22}$$

That is,

if $\Phi = \pi/2$,   if $\delta > -0.014544$, $\omega_{sunset} = \pi$, otherwise $\omega_{sunset} = 0$

if $\Phi = -\pi/2$,   if $\delta > -0.014544$, $\omega_{sunset} = 0$, otherwise $\omega_{sunset} = \pi$

if $\left[\sin(-0.014544) - \sin(\Phi)\sin(\delta)\right]/\left(\cos(\Phi)\cos(\delta)\right) \geq 1$, $\omega_{sunset} = 0$ $\qquad$ (2.23)

if $\left[\sin(-0.014544) - \sin(\Phi)\sin(\delta)\right]/\left(\cos(\Phi)\cos(\delta)\right) \leq -1$, $\omega_{sunset} = \pi$

otherwise $\omega_{sunset} = \cos^{-1}\left(\left[\sin(-0.014544) - \sin(\Phi)\sin(\delta)\right]/\left(\cos(\Phi)\cos(\delta)\right)\right)$

where $\cos^{-1}()$ is the arc cosine function. Once ω_{sunset} is obtained, $\omega_{sunrise}$ is equal to $-\omega_{sunset}$, and the above equations yield the times at sunrise and sunset.

If there are obstructions on the horizon, such as hills, mountains, buildings, or vegetation, this should be taken into account to obtain realistic times of sunrise and sunset. Assume that the angular elevations of the obstacles are known and that they are denoted $\gamma_{sunrise\_obstacle}$ and $\gamma_{sunset\_obstacle}$, respectively, in the directions of rising and setting for the day of interest. Sunrise occurs when the elevation angle of the sun γ_S is equal to $\gamma_{sunrise\_obstacle}$, or ($\gamma_{sunrise\_obstacle} - 0.01454$ rad) if the effects of refraction of the solar rays by the atmosphere and the angular dimension of the solar disk are taken into account. The hour angles $\omega_{sunrise}$ and ω_{sunset} are the solutions of the following equations:

$$\cos(\Phi)\cos(\delta)\cos(-\omega_{sunrise}) = \sin(\gamma_{sunrise\_obstacle}) - \sin(\Phi)\sin(\delta) \qquad (2.24)$$

$$\cos(\Phi)\cos(\delta)\cos(\omega_{sunset}) = \sin(\gamma_{sunset\_obstacle}) - \sin(\Phi)\sin(\delta)$$

2.4 Daytime – daylength

Daytime, or daylength, is defined as the period of time during which the sun is above the horizon during a day of 24 h. Daytime is noted S_{day} and is given in h by:

$$S_{day} = (t_{sunset} - t_{sunrise}) = (12/\pi)(\omega_{sunset} - \omega_{sunrise}) \qquad (2.25)$$

where ω_{sunset} and $\omega_{sunrise}$ are the hour angles at sunset and sunrise, in rad. In the absence of obstacle on the horizon, $\omega_{sunrise} = -\omega_{sunset}$. In this case, daytime is also called astronomical daytime and is noted S_0. It comes:

$$S_0 = (24\omega_{sunset}/\pi) = -(24\omega_{sunrise}/\pi) \qquad (2.26)$$

Figure 2.7 depicts the astronomical daytime S_0 as a function of the number of the day in the year at several latitudes. It may be noted that the shape of the variations in S_0 is the same than that in ω_{sunset}, and is therefore opposed to that of $\omega_{sunrise}$ (Figure 2.6) because S_0 is equal to the product of ω_{sunset} by a constant (equation 2.26).

The graph shows symmetry between north and south. Daytime S_0 is always equal to 12 h at equator. It is also equal to 12 h at any latitude during the days of equinoxes (approximately days #81 and #265), except at poles. In the northern hemisphere, the daytime is greater during the period between equinoxes of March and September

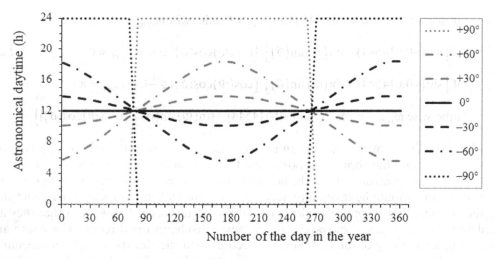

Figure 2.7 Astronomical daytime S_0 as a function of the number of the day in the year at several latitudes.

Table 2.1 Yearly average of the astronomical daytime S_0 at various latitudes

Latitude	Astronomical daytime (h)	Latitude	Astronomical daytime (h)
90°	12.23 (or 12 h 18 min)	−90°	11.77 (or 11 h 46 min)
60°	12.10 (12 h 6 min)	−60°	11.90 (11 h 54 min)
45°	12.06 (12 h 4 min)	−45°	11.94 (11 h 56 min)
30°	12.03 (12 h 2 min)	−30°	11.97 (11 h 58 min)
0°	12.00 (12 h)		

than between September and March. It is the opposite in the southern hemisphere. The maximum in daytime is reached at the summer solstice (June in the northern hemisphere and December in the southern one), whereas the minimum is attained at the winter solstice. S_0 exhibits variations during the year, whose amplitude increases as one move away from the equator. At latitudes −30° and 30°, the daytime varies between approximately 10 h 5 min and 13 h 55 min. It fluctuates between 5 h 34 min and 18 h 26 min at latitudes −60° and 60°. At the poles, the daytime is either 24 or 0 h.

Figure 2.7 is of practical interest. Knowing that the daytime is centered at 12:00, it is possible to estimate the times at sunrise and sunset in TST. For example, at equator, the daytime is 12 h. Hence, sunrise occurs at 06:00 (= 12:00−12/2) and sunset at 18:00 (= 12:00 + 12/2).

Table 2.1 gives the yearly average of the daytime S_0 at various latitudes. The average is close to 12 h everywhere though a dissymmetry between the northern and southern hemispheres may be noted: The daytime is greater in the former than in the latter. The average is 12 h at equator. The maximum is 12.23 h at the North Pole. The average decreases with latitude and reaches a minimum of 11.77 h at the South Pole. The difference between these extrema is 0.46 h, or about 28 min.

If the angular dimension of the solar disk and the effects of the refraction of solar rays by the atmosphere are accounted for, then the daytime is greater than that mentioned above. Use equation (2.22) to calculate the hour angle ω_{sunset} at sunset. For example, at the equator, the daytime will vary slightly during the year depending on the solar declination δ. ω_{sunset} is given by:

$$\sin(-0.014544) = \cos(\delta)\cos(\omega_{sunset}) \tag{2.27}$$

Thus, during the equinoxes, for which $\sin(\delta) = 0$ and $\cos(\delta) = 1$ in equation (2.22), the daytime is about 12 h 7 min, and the night lasts 11 h 53 min at equator, and no longer 12 h. At latitude 60°, the daytime is about 12 h 13 min, and not 12 h.

2.5 Solar angles during equinoxes and solstices

Figures 2.8–2.13 show the solar zenithal angles θ_S and azimuths Ψ_S during the particular days of equinoxes and solstices at various latitudes, from equator to the poles.

Why this choice of days and latitudes? During the year, at a given site, the course of the sun varies between the curves shown in these graphs which exhibit the extrema.

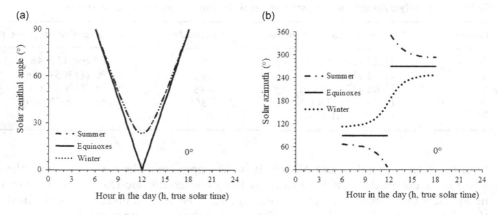

Figure 2.8 Solar zenithal angle θ_S (a) and azimuth Ψ_S (b) during days of solstices and equinoxes at equator. Curves are not drawn for hours when the sun is below the horizon ($\theta_S > 90°$).

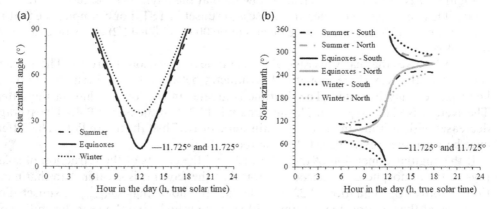

Figure 2.9 Solar zenithal angle θ_S (a) and azimuth Ψ_S (b) during days of solstices and equinoxes at latitudes $-11.725°$ and $11.725°$. Curves are not drawn for hours when the sun is below the horizon ($\theta_S > 90°$).

In other words, the angles θ_S and Ψ_S vary between the curves at equinoxes and those at solstices during the year. For a given day, the angles vary either between the curves shown here at the equator and the tropics or between those at the tropics and the poles depending on the latitude. Hence, these curves are a means to make a rough estimate of θ_S and Ψ_S for other days and at other latitudes.

Reader, you may find on the Web nicer graphs in pseudo-3D showing the course of the sun during these particular days and the changes in solar angles. I made these graphs on purpose to offer an accurate reading of both θ_S and Ψ_S. As they are extreme values of these angles, such graphs allow a quick assessment of the behavior of these angles during the year and in space, as a first approximation, or a gross verification of the angles obtained by an algorithm or a spreadsheet.

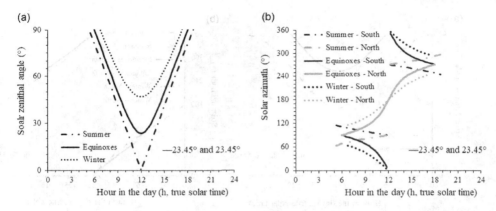

Figure 2.10 Solar zenithal angle θ_S (a) and azimuth Ψ_S (b) during days of solstices and equinoxes at latitudes −23.45° and 23.45°. Curves are not drawn for hours when the sun is below the horizon ($\theta_S > 90°$).

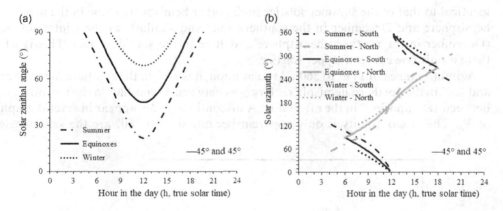

Figure 2.11 Solar zenithal angle θ_S (a) and azimuth Ψ_S (b) during days of solstices and equinoxes at latitudes −45° and 45°. Curves are not drawn for hours when the sun is below the horizon ($\theta_S > 90°$).

In Figures 2.8–2.13, the words *Summer – South*, respectively *Winter – South*, mean the summer solstice (December) and winter solstice (June) in the southern hemisphere. Similarly, *Summer – North*, respectively *Winter – North*, mean the summer solstice (June) and winter solstice (December) in the northern hemisphere. The words *Equinoxes – North* and *Equinoxes – South* mean the March and September equinoxes in the northern and southern hemispheres, respectively.

The leftmost plots are for the solar zenithal angle and the rightmost ones for the azimuth. Curves are not drawn for hours when the sun is below the horizon ($\theta_S > 90°$). The curves for solar zenithal angles are identical from one equinox to the other. Hence, there is only one curve for θ_S for equinoxes, regardless of the hemisphere. Finally, there is a symmetry in astronomical seasons between the northern and southern hemispheres for θ_S: The solar zenithal angle at the summer solstice in an hemisphere is

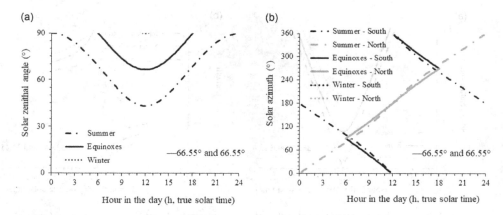

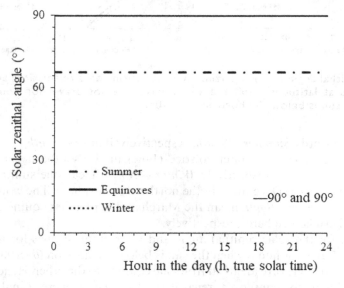

Figure 2.12 Solar zenithal angle θ_S (a) and azimuth Ψ_S (b) during days of solstices and equinoxes at latitudes $-66.55°$ and $66.55°$ (polar circles). Curves are not drawn for hours when the sun is below the horizon ($\theta_S > 90°$).

identical to that of the summer solstice in the other hemisphere (June in the northern hemisphere and December in the southern one) and similarly at the winter solstice (December in the northern hemisphere and June in the southern one). This is why there is only one curve of θ_S per solstice.

Azimuth ranges from 0° to 360°. At solar noon, it is zero in the southern hemisphere and 180° in the northern hemisphere. It ranges between 0° and 180° in the morning and between 180° and 360° in the afternoon. A discontinuity may appear in several graphs of Ψ_S. This discontinuity is only apparent because 0° and 360° are the same angle.

Figure 2.13 Solar zenithal angle θ_S during days of solstices and equinoxes at poles. Curves are not drawn for hours when the sun is below the horizon ($\theta_S > 90°$).

I recall that the azimuth cannot be calculated at the poles. Hence, there is no plot for azimuth in Figure 2.13 that deals with the poles.

Figure 2.8 deals with the equator. During the equinoxes, the sun has an apparent vertical movement (Figure 2.8a): The solar zenithal angle decreases linearly from sunrise until noon to reach the value 0° (linear increase in solar elevation angle until maximum of 90°), then increases likewise until sunset (decrease in elevation angle to a minimum of 0°). The sun rises in the east ($\Psi_S = 90°$) and its azimuth remains constant until noon (Figure 2.8b), when the sun passes at the zenith of the observer ($\theta_S = 0$). Then, when the sun goes down on the horizon, its azimuth remains constant and equal to 270° ($3\pi/2$).

The θ_S curves for the two solstices are superimposed: There is no difference. However, the azimuth curves differ. At the summer solstice (June), the sun rises east–northeast ($\Psi_S = 67°$). Its course is bent to the north with an azimuth equal to 0° at noon, then to the west, and the sun sets to the west–northwest. At the December solstice (winter in the figure), the sun rises east–southeast ($\Psi_S = 113°$), passes south ($\Psi_S = 180°$), and sets in west–southwest.

An equatorial zone may be defined which is comprised between latitudes −11.725° and +11.725°. It is the central half of the intertropical zone which is comprised between latitudes −23.45° and +23.45° (11.725 = 23.45/2). Changes in θ_S and Ψ_S in this equatorial zone are plotted in Figure 2.9. θ_S reaches its minimum at equinoxes when the sun is at its highest during the year. θ_S is greater in winter than in summer for the same hour; i.e., the sun is closer to the horizon. At solar noon, the sun is as high at summer solstice as at equinoxes ($\theta_S = 11.725°$; Figure 2.9) at latitudes −11.725° and +11.725°.

In the equatorial zone, at the equinoxes, the sun rises in the east ($\Psi_S = 90°$) and sets in the west ($\Psi_S = 270°$) after passing north (360°) if in the southern hemisphere, south (180°) if not. At the summer solstice in the southern hemisphere, or at the winter solstice in the northern hemisphere, the sun rises in the east–southeast (113°), goes south (180°), and sets in the west–southwest (247°) without the Ψ_S curves overlapping. At the winter solstice in the southern hemisphere, or at the summer solstice in the northern hemisphere, the sun rises east–northeast (67°). Its course turns to the north with an azimuth equal to 0° at noon; then the sun sets toward the west–northwest (293°).

Outside this equatorial zone, the apparent movement of the sun in the sky takes the form of a bell (Figures 2.10–2.12), except at the poles which form another special case.

For the same true solar time, θ_S reaches its minimum during the summer solstice and its maximum during that of winter. In other words, the sun is higher in summer than in winter at the same time. At the equinoxes, the azimuth is equal to 90° (east) at sunrise and 270° (west) at sunset. In winter, the azimuth is greater than 90° at sunrise and less than 270° at sunset. On the contrary, in summer, the azimuth is less than 90° at sunrise and greater than 270° at sunset. Between the tropics and the poles, the azimuth at solar noon is equal to 180° in the northern hemisphere and is zero in the southern hemisphere, except at the poles themselves where there is indetermination.

Figure 2.10 exhibits the case of the tropics of Capricorn (−23.45°) and Cancer (23.45°), which obeys the general case described above, except at the summer solstice. During this day, the sun has an apparent vertical movement (Figure 2.10a): The solar zenithal angle decreases linearly from sunrise until noon to reach the value 0, then increases in the same way until sunset. In the southern hemisphere, the sun rises east–southeast ($\Psi_S = 113°$). Its course bends north until noon (Figure 2.10b), when the sun passes at the zenith of the observer ($\theta_S = 0$); then, the sun sets to the west–southwest (247°). In the northern hemisphere, the sun rises east–northeast ($\Psi_S = 67°$). Its course bends south until noon, when the sun passes over the zenith of the observer ($\theta_S = 0$); then, the sun sets to the west–northwest ($\Psi_S = 293°$).

The general case is illustrated in Figure 2.11 depicting changes in θ_S and Ψ_S at latitudes −45° and 45°.

Figure 2.12 deals with the case of the polar circles (−66.55° and 66.55°). It obeys the general case, except at winter solstice during which the sun touches the rim of the horizon at solar noon. Azimuth cannot be calculated for this day. Correspondingly, the sun is always visible during the day at the summer solstice. It is near the horizon at the beginning of the day, and its elevation angle reaches a maximum of 47° ($\theta_S = 43°$) at solar noon. The azimuth ranges from 0° to 360° in the southern hemisphere and from 180° to 0° and then 360° to 180° in the northern hemisphere. The sun revolves around the observer in its apparent motion.

At latitudes closer to the poles, the sun is below the horizon during a period of several days, centered at around the winter solstice. The closer you are to the poles, the longer the polar night. Figure 2.13 exhibits the variations in θ_S at poles. The curve at the winter solstice cannot be seen because the sun is below the horizon ($\theta_S > 90°$). At equinoxes, the sun touches the rim of the horizon ($\theta_S = 90°$) and the elevation angle is zero. The sun is always above the horizon ($\theta_S < 90°$) in the period between the equinoxes of spring and autumn. The maximum of θ_S is reached at summer solstice, when the solar zenithal angle is constant during the day and equal to 66.55° (elevation angle of 23.45°).

2.6 Effective solar angles

The equations above show that at a given location and a given day, the solar angles depend entirely on the true solar time. Since the radiation measurements are made over a certain period of time during which the true solar time and therefore the solar angles change, what are the solar angles to be assigned to each measurement? Measurements or estimates of solar radiation are integrated over a certain time, called the integration period or summarization. Therefore, the above equations cannot be applied as such.

Why should you care? It is very often necessary to assign solar angles to each measurement. An example of such a need is the computation of the corresponding extra-terrestrial radiation which is necessary to control the quality of measurements and should be known for the integration period. Another example is the computation of the angle θ under which the sun is seen from an inclined surface which is the major factor in the calculation of the received solar radiation. In the two examples, a solution consists in computing the radiation every 30 s or 1 min within the integration period, then summing the results. Is there a simpler and faster way of doing it?

In cases where the integration period is 1 min or less, then it can be considered with good accuracy that the solar angles are constant over this duration. During such short periods, the angles may be considered as varying linearly and they can be calculated for the time in the middle of the period. For example, if the measuring interval is between 09:25 and 09:26, angles can be calculated at 09:25:30.

This is no longer the case when the integration period is greater than 1 min. The angles vary greatly during such a period, in particular at the time of sunrise or sunset. In this context, how to calculate the corresponding solar zenithal angle θ_S and azimuth Ψ_S? The concept of effective angles is defined to answer this practical problem. The effective solar angles are fictitious and correspond as best as possible to the solar radiation measured during the period. The effective solar angles are noted θ_S^{eff} and Ψ_S^{eff}.

2.6.1 Methods for computing effective angles

The problem comes down to the assessment of the effective solar zenithal angle θ_S^{eff}. Once θ_S^{eff} is estimated, the effective solar azimuth Ψ_S^{eff} can be deduced by the equations already presented. I do not know of a general solution to this daily problem. Philippe Blanc and I did a fairly simple study in 2016 by looking at some common methods and suggesting others. The resulting publication[2] does not provide a definitive answer to this question, but provides some concrete, easy-to-implement solutions, which are described below.

The concept of the 2016 study is based on the assessment of the direct component of the solar radiation that will be detailed in the following chapters. For now, I just have to say that this component, noted B, is the part of the solar radiation that appears to come from the sun. It therefore excludes the other part of the solar radiation coming from the other directions of the sky vault and called diffuse component. Because it appears to come from the direction of the sun, the angle of incidence of B onto a horizontal plane is equal to the solar zenithal angle θ_S^{eff}. The direct component B is related to the direct component B_N at normal incidence by the following equation:

$$B = B_N \cos\left(\theta_S^{eff}\right) \tag{2.28}$$

In this study, Philippe and I used very accurate 1-min measurements of the direct component B_N at normal incidence available at two measuring stations, from which we computed 1-min B. Then, for the sake of simplicity, we assumed hourly measurements, i.e., with an integration period Δt of 1 h. We thus computed hourly averages of B_N and B. Each hourly average of B was assumed to be a measurement for which θ_S^{eff} must be computed by one of the tested methods. Finally, the hourly B_N was estimated by the equation

$$B_N = B \cos^{-1}\left(\theta_S^{eff}\right) \tag{2.29}$$

where $\cos^{-1}()$ is the arc cosine function and was compared to the actual value.

Three common methods were tested first. Let t be the time at the end of the integration period, expressed in h. In Method #1, θ_S^{eff} is equal to θ_S at half-hour, i.e., at $(t-0.5)$:

$$\theta_S^{eff} = \theta_S \operatorname{at}(t-0.5) \tag{2.30}$$

2 Blanc P., Wald L., 2016. On the effective solar zenith and azimuth angles to use with measurements of hourly irradiation. *Advances in Science and Research*, 13, 1–6, doi:10.5194/asr-1-1-2016.

In Method #2, θ_S^{eff} is equal to the average of θ_S over the hour, i.e., between $(t-1)$ and t:

$$\theta_S^{eff} = (1/\Delta t)\int_{t-1}^{t}\theta_S(u)du \tag{2.31}$$

In Method #3, θ_S^{eff} is equal to the average of θ_S over the hour, with limitations to the times of sunrise and sunset:

$$\theta_S^{eff} = \int_{t-1}^{t}\theta_S(u)du \Big/ \int_{t-1}^{t} du, \quad \text{only for } u \text{ such as } \theta_S(u) < \pi/2 \tag{2.32}$$

Comparisons with actual measurements showed that the relative error on the solar radiation (i.e., B_N) depends on θ_S. The closer the sun to the horizon, the greater the error. In any case are the errors negligible. They are the greatest for Methods #1 and #2. The root mean square error is about 30 %–40 % relative to the average radiation. The bias is the average of the errors and denotes a systematic error. It is of a few percent in relative value; it may reach 16 % when $\theta_S > 75°$. The best results are given by Method #3. The relative root mean square error is about 4 %–5 %, though it may reach 24 % when $\theta_S > 75°$. The relative bias is small, except when $\theta_S > 75°$, where it may reach 23 %.

Two novels methods were proposed by Philippe and I, where θ_S^{eff} is not computed from averages of θ_S. Taking advantage of equation (2.28), θ_S^{eff} is given by models computing the direct components B and B_N over the hour for the extraterrestrial radiation (Method #4) and cloudless atmosphere (Method #5). In Method #4, the hourly extraterrestrial radiation B_0 and B_{0N} are computed, and their ratio yields θ_S^{eff}:

$$\theta_S^{eff} = \cos^{-1}(B_0/B_{0N}) \tag{2.33}$$

In Method #5, the hourly radiation B_{clear} and B_{Nclear} for cloudless conditions are computed by a so-called clear-sky model, and their ratio yields θ_S^{eff}:

$$\theta_S^{eff} = \cos^{-1}(B_{clear}/B_{Nclear}) \tag{2.34}$$

Method #4 provides similar errors to those of Method #3. The smallest errors are obtained by Method #5, for which the bias is almost negligible, whatever θ_S. The relative root mean square error is 2 %, except when $\theta_S > 75°$ where it may reach 9 %.

2.6.2 How to practically use one of these methods

Which lessons can be learnt from this study? Philippe and I recommend not using Methods #1 and #2. The errors are too large and include an important bias that may reveal troublesome in subsequent calculations. Methods #3 and #4 offer smaller but still noticeable errors. This should be accounted for in subsequent calculations. Errors tend to dramatically increase when the sun is low on the horizon, at sunrise and sunset. Method #5 gives the best results by far, with small bias and root mean square error. It shows a limited degradation of performance when θ_S increases.

How to practically use one of these methods? Methods #1–3 need the calculation of θ_S every minute. To do so, one may use the equations given above, or already mentioned software libraries. One may also use Web services such as the SoDa Service specialized in solar radiation (www.soda-pro.com). This site proposes services to which you can provide inputs such as beginning and end dates or geographic coordinates, and that return time series of solar angles for various integration periods in tabulated form. They can be copied and pasted in a spreadsheet for further use.

Method #4 requests estimates of the solar radiation at the top of the atmosphere on a horizontal plane and at normal incidence. They can be calculated by means of the equations given in the following chapter, or by means of Web services, including those available in the SoDa Service.

Of course, it is best to use Method #5 because it is the most accurate. It requires that a clear-sky model is available that provides estimates of B_{clear} and B_{Nclear} in cloud-free conditions. You may wish to implement one yourself. The book by Daryl Myers documents several such models[3], and several codes are available on the Web.[4] Then, you have to find sources of data about the clear atmosphere such as the water vapor content, the Linke turbidity factor or the aerosol optical depth as it will be seen in the following chapters. These data are input to the clear-sky model. The direct component of the radiation is computed every minute within the integration period for the horizontal plane and at normal incidence. Then, the results are summed yielding B_{clear} and B_{Nclear}, and finally θ_S^{eff}.

Another practical means for exploiting Method #5 without writing a code dedicated to a clear-sky model and obtaining the necessary inputs to the model is to use online services. One example is the McClear application in the SoDa Service, also part of the Radiation Service of the European Copernicus Atmosphere Monitoring Service (CAMS). It exploits the McClear clear-sky model and returns time series of irradiances, including B_{clear} and B_{Nclear}. Users must enter the place of interest or geographic coordinates, desired period of time, time system, and integration period from 1 min to 1 month. The application itself reads the database of estimates of constituents of the cloudless atmosphere carried out daily by the CAMS. Connoisseurs may be interested to know that the McClear application is a Web processing service, a standard from the Open Geospatial Consortium (OGC). It can be invoked automatically from a computer code written, for example, in Python or other language. It thus offers an opportunity to fully automate the process of computing effective angles.

Methods #4 and #5 call upon models that are often available for the total radiation, i.e., integrated over the whole solar spectrum. What should you do if you deal with other spectral ranges such as the UV or visible? Nothing more. These models for total radiation are suitable in these cases as the solar angles do not depend on the wavelength. The effective solar zenithal angle θ_S^{eff} is computed from the ratio B_{clear}/B_{Nclear}, which is independent of the wavelength as shown by equation (2.28).

3 Myers D.R., 2017. *Solar Radiation: Practical Modeling for Renewable Energy Applications*, Boca Raton: CRC Press.

4 https://github.com/EDMANSolar/pcsol, written by Oscar Perpiñán Lamigueiro. [The author gives R codes for several tens of clear-sky models.]

Chapter 3

Solar radiation incident at the top of the atmosphere

At a glance

The radiation originating from the sun received at the top of the atmosphere is also called the extraterrestrial radiation. It varies over time due to changes in the position of the Earth relative to the sun and, to a lesser extent, variations in solar activity from day to day. The closer to the Earth the sun, the greater the extraterrestrial radiation.

The total solar irradiance E_{TSI} is the yearly average of the extraterrestrial irradiance received on a plane located at the top of the atmosphere at normal incidence and integrated over the whole spectrum. It varies very little from year to year, about 1 W m^{-2}. The International Astronomical Union recommends a value of 1361 ± 1 W m^{-2} for the period May 1996–January 2008, known as solar cycle #23. The value 1362 W m^{-2} has been measured in 2010. The relative change of E_{TSI} during a year is comprised between +3 % and −3 %. The inter-daily variations may reach 5 W m^{-2} that is about 0.4 % of E_{TSI}.

The irradiance received by a horizontal surface on the ground is always less than its equivalent received at the top of the atmosphere, except in special cloudy cases.

The spectral distribution of the extraterrestrial irradiance exhibits a peak around 500 nm. About half of the irradiance is in the visible and near-infrared range of the electromagnetic spectrum. The spectral interval is not wide: About 98 % of the extraterrestrial irradiance is between 0.3 and 4 µm, and 99.9 % is between 0.2 and 8 µm. The radiation integrated over the whole spectrum is called total radiation (total irradiance, total irradiation). The spectral distribution of the extraterrestrial irradiance varies with time. The irradiance in the ultraviolet range is the most variable, while that in the visible or infrared range is less.

Solar radiation from the sun received at the top of the atmosphere, about 120 km above sea level, is also called extraterrestrial radiation. Its accurate knowledge is necessary for the estimation of the incident radiation on the ground. Extraterrestrial radiation is attenuated by the atmosphere in its path toward the ground. Consequently, the value at the top of the atmosphere constitutes an upper bound that the radiation received by a horizontal surface on the ground will not reach, except in some special cases of multiple reflections on clouds.

A few radiometric quantities must be defined to properly describe the radiation, notably the radiance, the irradiance, and the irradiation. This is the first part of this chapter. These quantities can also be defined at each wavelength λ and are then called spectral quantities. Rules for converting irradiance to irradiation and vice versa are given. Note that the usual notation for the wavelength is λ, which should not be confused with the longitude.

After having recalled the variations of the sun–Earth distance and by briefly presenting the solar activity, I give the means to estimate the quantity of energy received on a horizontal plane, or inclined plane, at the top of the atmosphere for a place, a date, and a given time. I first deal with the case of so-called total radiation, i.e., the radiation integrated over all wavelengths. Then, I describe the spectral distribution of solar radiation at the top of the atmosphere.

3.1 Radiance – irradiance – irradiation

I define here several quantities necessary for the description of the radiation, whether solar radiation or another, at the top of the atmosphere or on the ground or any altitude. The radiometric quantities defined below can be defined at any wavelength λ; they are then called spectral quantities. They can also be used for the whole spectrum, in which case they are called total quantities.

To avoid confusion with photometry that deals with radiation from the point of view of human vision, the quantities below are often qualified as energetic quantities and are part of radiometry. For example, I could write energetic radiance, energetic irradiance, etc. Once this is posed, I allow myself not to use the energetic adjective in the following to simplify the writing.

The radiant energy Q is the quantity of energy transferred by radiation. It is expressed in joule (J). The power is an energy divided by a time duration Δt. The radiant flux F is a radiant power expressed in watt (W); it is an instantaneous value of a radiant energy:

$$F = Q/\Delta t \tag{3.1}$$

In the case of an opaque collector plane not being located on the ground, the downward flux can be defined as the radiative flux received by the upper face of the plane and the upward flux as that received by the lower face.

3.1.1 Radiance

If the radiative source offers some extent, the radiance may be derived from the flux. The radiance is the quantity of flux per unit of solid angle received per unit of surface. It is expressed in W m^{-2} sr^{-1}. Figure 3.1 is a schematic view of the definition of the radiance. If dF denotes the flux element originating from a source seen from a solid angle $d\Omega$ from the receiving surface of area dS and if θ denotes the angle between the normal to the surface dS and the beam, the radiance $L(\theta)$ is given by:

$$d^2F = L(\theta)\,dS\cos(\theta)\,d\Omega \tag{3.2}$$

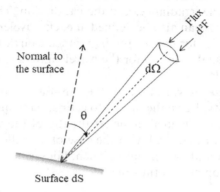

Figure 3.1 Definition of the radiance.

The radiance is a function of the direction of illumination of the collecting surface. The diagram describing the variation in radiance as a function of this direction is called a radiation pattern or diagram. When the radiance is independent of the direction of illumination, that is to say, isotropic, the pattern is a hemisphere and $\Omega = 2\pi$.

The spectral radiance is the quantity of flux per unit of solid angle received per unit of surface at a given wavelength λ and is expressed in W m^{-2} sr^{-1} nm^{-1}:

$$L(\lambda,\theta) = d^2F(\lambda)/(dS\cos(\theta)d\Omega) \tag{3.3}$$

The angle θ is called the angle of incidence. The radiance L decreases when the angle of incidence θ increases. When θ is equal to 0, incidence is said normal, because in geometry, perpendicular to the surface is also called normal to the surface. Radiance L is then at its maximum. When $\theta = \pi/2$, the incidence is said grazing and the radiance is minimum and equal to 0.

3.1.2 Irradiance

The irradiance is the integral of the radiances over the hemisphere above the collector plane. Let θ and ζ be the angles describing the direction of illumination, and the irradiance E is given by:

$$E = dF/dS = \int_{\text{hemisphere}} L(\theta,\zeta)\cos(\theta)d\Omega \tag{3.4}$$

The irradiance is a density of radiant flux, or in other words, the power received per unit area. It is expressed in watt per square meter (W m^{-2}). The spectral irradiance $E(\lambda)$ is the irradiance at a given wavelength and is expressed in W m^{-2} nm^{-1}. The total irradiance is the irradiance integrated over the whole spectrum of the solar radiation.

If the opaque collector plane is above the ground, the upwelling irradiance is defined as the integral of the upwelling radiances over the hemisphere located below the collector.

Conversely, the radiance is the hemispherical distribution of the irradiance. If the hemispherical distribution is uniform or orthotropic, that is, if the radiance is angularly uniform and does not depend on the direction of illumination (θ, ζ), then there is a simple relationship between E and L:

$$E = L = \int_{\text{hemisphere}} \cos(\theta)\,d\Omega = \pi L \tag{3.5}$$

In the case of the solar radiation at the top of the atmosphere, the distribution of the radiance is not orthotropic. The only source in the hemisphere above the collecting surface is the sun, which is seen from a solid angle of $0.68 \ 10^{-4}$ sr, with an apparent diameter of approximately $32' \pm 0.5'$ of arc. The extraterrestrial irradiance, or the irradiance at the top of the atmosphere, is therefore only due to the integration of the radiances on the solid angle under which the sun is seen. This contribution is called the direct component of the irradiance.

3.1.3 Irradiation – radiant exposure

The irradiation H is a density of radiant energy, or in other words the energy received per unit area during a given duration. It is expressed in J m^{-2} (joule per square meter), or J cm^{-2}, or kJ m^{-2}. The spectral irradiation $H(\lambda)$ is the irradiation at a given wavelength and is expressed in J m^{-2} nm^{-1}. The total irradiation is the irradiation integrated over the whole spectrum of the solar radiation.

The irradiation H is the integral of the irradiance E for a given duration Δt:

$$H(t, \Delta t) = \int_{t-\Delta t}^{t} E(u)\,du \tag{3.6}$$

I insist on a "given duration". Let me take an example to illustrate the importance of the duration. Assume a constant irradiance E over time. If I measure during $\Delta t = 1$ h, I will get a certain value of irradiation $H_1 = E \ \Delta t$. If I measure during 2 h, I will get a second measurement $H_2 = 2 \ E \ \Delta t$ which will be twice the first. So always mention the duration of the measurement. For example, write hourly irradiation, respectively daily irradiation, to describe the energy density measured during 1 h, 1 day.

The duration of measurement is also called summarization, or integration duration, or integration period, or integration time, or measurement period.

Irradiation is what is measured in meteorological networks and many other networks. It is measured on a horizontal plane, except for the direct component of the irradiation which is measured at normal incidence as it will be seen later. The daily irradiation is the density of energy collected on a horizontal plane between the sunrise and the sunset, since the solar radiation received at ground is zero during the night. The monthly irradiation, respectively yearly or annual irradiation, is the density of energy received during a month, a year.

In its Guide to Meteorological Instruments and Methods of Observation,[1] the World Meteorological Organization dos not use the term irradiation but the terms radiant

---

1 Guide to Instruments and Methods of Observation. WMO-No. 8, 2018 edition, 2018, World Meteorological Organization, Geneva, Switzerland. [Vol 1 Measurement of Meteorological Variables, Chapter 7, Measurement of Radiation.]

exposure, spectral exposure, and radiant exposure for 1 h or for a day, or hourly or daily sums, or totals, of radiant exposure instead.

Practitioners in the field of electricity production use another unit for irradiation: Wh m^{-2} (watt-hour per square meter). Since 1 h = 3600 s and 1 J = 1 W s, the conversion factors are:

$$1\,\text{Wh m}^{-2} = 3600\,\text{W s m}^{-2} = 3600\,\text{J m}^{-2} = 0.36\,\text{J cm}^{-2} = 3.6\,\text{kJ m}^{-2} \tag{3.7}$$

Reader, I draw your attention to the unit used in documents, reports, maps, and data, in the field of energy, by federal agencies of the United States, such as the laboratory for renewable energies NREL, or the aerospace agency NASA. These agencies are required by federal law to use the unit Wh m^{-2} day^{-1} to express daily irradiation, including monthly or annual averages. This "political" unit is not that of an irradiation since it is divided by a time.

Another unit of measurement, the langley (Ly), is sometimes used in the field of aging of materials, paints, and coatings. One langley is equal to 41.84 kJ m^{-2} or 0.04184 MJ m^{-2}.

3.1.4 Converting irradiance to irradiation and vice versa

How to convert irradiance in irradiation and vice versa? Given a measurement of irradiation H and knowing that the irradiance varies during the duration of measurement Δt, the average irradiance E is deduced from H by:

$$E(t) = H(t, \Delta t)/\Delta t \tag{3.8}$$

Like irradiation, the duration of measurement Δt must be given for irradiance. For example, you must write *hourly average of solar irradiance* to describe the averaged irradiance E computed from the hourly irradiation.

The hourly average of irradiance in W m^{-2} is converted to hourly irradiation by multiplying it by 3600 s, i.e., the number of seconds in 1 h. The result is given in J m^{-2} because 1 J = 1 W s. The hourly irradiation expressed in Wh m^{-2} has the same value than the hourly average of irradiance in W m^{-2}. Conversely, the hourly irradiation in J m^{-2} is converted to hourly average of irradiance in W m^{-2} by dividing it by 3600 s.

Let me take another example: The irradiance averaged over 10 min, in W m^{-2}, is converted to 10-min irradiation expressed in J m^{-2}, by multiplying it by 600 s, i.e., the number of seconds in 10 min. The irradiation during 10 min, but expressed in Wh m^{-2}, is obtained by dividing the irradiance averaged over 10 min by 6, i.e. the number of times one finds 10 min in 1 h.

The case of the timescales greater than or equal to 1 day is a little different and is governed by a convention in the climate field. The daily average of irradiance in W m^{-2}, is equal to the daily irradiation in J m^{-2} divided by the number of seconds in 24 h, or 86,400 s. By doing so, daily averages of irradiance may be compared with each other without taking into account the daytime. For example, a daily irradiation of 1 MJ m^{-2} (10^6 J m^{-2}) is equivalent to a daily average of irradiance of 11.6 W m^{-2}. If the daily irradiation is expressed in Wh m^{-2}, the daily average of irradiance is obtained by dividing it by 24 h.

The calculation of the monthly and annual irradiations, and their averages as well as the corresponding irradiances, is detailed in Chapter 9, which addresses also the case of the incomplete series of measurements and the specific recommendations of the World Meteorological Organization.

3.2 Solar activity – extraterrestrial radiation

The sun is the seat of thermonuclear phenomena producing a vast amount of energy. Part of this energy is emitted in the form of radiation, called, unsurprisingly, solar radiation. Energy is equally radiated in all directions and decreases with the square of the distance between the sun and the receiver. Despite the very long distance between the sun and the Earth, the radiated energy reaching the Earth is extremely high. It is the main source of natural energy on the Earth, far ahead of the other sources which are the geothermal heat flux generated from inside the Earth, natural terrestrial radioactivity, and cosmic radiation. These other sources produce about 0.1 % of the total amount and are all negligible compared to solar radiation.

The sun is a very active star. Observation of its surface temperature shows that it is not spatially uniform and that it varies in time. Sunspots appear whose temperatures are less than those of the surrounding areas. Solar flares and coronal mass ejections may also be observed. The ensemble of these events is called solar activity. The solar activity exhibits minima and maxima that are characterized by the number of sunspots. The succession of minima and maxima defines the solar cycle whose period varies between 8 and 15 years, with an average of about 11 years. According to the International Astronomical Union, the first cycle was listed around the year 1760. The cycle #24 started in 2008 and may end in late 2020.

The radiated energy reaching the top of the atmosphere of the Earth is also called the extraterrestrial radiation. It varies over the course of the year due to changes in the position of the Earth relative to the sun and, to a lesser extent, intra- or inter-day variations in solar activity. Let $E_{0N}(t)$ be the extraterrestrial irradiance incident on a plane located at the top of the atmosphere and always normal to the direction of the sun, therefore with a zero angle of incidence, at a given time t. $E_{0N}(t)$ depends on the distance $r(t)$ between the sun and the Earth; the closer the Earth to the sun, the greater $E_{0N}(t)$. If r_0 denotes the distance averaged over a year, the change in $E_{0N}(t)$ around its annual average E_{TSI} is a function of the square of the ratio $(r_0/r(t))$ and $E_{0N}(t)$ is given by:

$$E_{0N}(t) = E_{TSI}\left[r_0/r(t)\right]^2 \tag{3.9}$$

But what is E_{TSI}? E_{TSI} is called total solar irradiance by astronomers and often abbreviated as TSI. This term may be confusing in our context because it makes no reference to the fact that it is a quantity at the top of the atmosphere and not at ground level. However, it is used in many documents and I adopt it in this book by adding the initials TSI to avoid confusion: I will call it the total solar irradiance TSI.

As written before, E_{TSI} is the annual average of $E_{0N}(t)$. For many years, E_{TSI} was called the solar constant because it was believed that it was exhibiting very small variations throughout the years. Measuring E_{TSI} with precision is difficult to do. Instrumentation and its use have evolved in recent decades with increased use of instruments on board satellites, starting in 1979. This development resulted in regular reviews of E_{TSI}.

The results depend on the radiometers on board the satellites, their calibration, and the knowledge of their aging. Several values of E_{TSI} are thus proposed in the scientific literature, which are close to one another. In 1981, the value recommended by the World Radiometric Reference was 1370 ± 6 W m^{-2}. The value 1367 W m^{-2} was often adopted in the year 2000. Recent measurements[2] made in 2010 give a value of 1362 W m^{-2} with an uncertainty of about 2 W m^{-2}. The International Astronomical Union recommends the value 1361 ± 1 W m^{-2} for the solar cycle #23, i.e., for the period May 1996–January 2008.[3] These different values are close to one another, and the influence of the solar cycle on E_{TSI} is very limited. The variations during a cycle are approximately 1 W m^{-2} or approximately 0.1 % in relative value. In comparison, the inter-daily variations in $E_{0N}(t)$ due to solar activity are greater and can reach 5 W m^{-2} or approximately 0.4 % of E_{TSI}.[4]

Considering only the variation of the sun–Earth distance, $E_{0N}(t)$ varies by a factor of approximately $2 \ 10^{-4}$ from 1 day to the next, that is about 0.2 W m^{-2}. This inter-daily variation is small, and the variation of E_{0N} from 1 h to the next is even smaller. Thus, it can be considered with sufficient accuracy that $E_{0N}(t)$ is constant over the day. In other words, it does not depend on the time t_{TST} or the hour angle ω nor consequently on the latitude Φ. Therefore, E_{0N} can be calculated by:

$$E_{0N}(t) = E_{TSI}\left(r_0/r(t)\right)^2 = E_{TSI}\left(1 + \varepsilon(t)\right) \tag{3.10}$$

where ε is the correction of relative eccentricity and is given by equation (1.6) as a function of the number of the day in the year. ε ranges from -0.03 to $+0.03$. In other words, $E_{0N}(t)$ ranges from 0.97 to 1.03 E_{TSI}. Taking E_{TSI} equal to 1361 W m^{-2}, $E_{0N}(t)$ is between 1315 and 1407 W m^{-2}.

Figure 3.2 shows the extraterrestrial irradiance $E_{0N}(t)$ as a function of the number of the day in the year 2005. The extraterrestrial irradiance at normal incidence is minimum when the distance r is the greatest, i.e., here on 4 July, i.e., day #185, and maximum when the distance is the smallest, i.e., on 3 January.

If the variations of the sun–Earth distance are neglected, i.e., if I write $E_{0N}(t) \approx E_{TSI}$, the error made on $E_{0N}(t)$ depends on the number of the day in the year and is between -45 and 45 W m^{-2}. The relative value of the error is between -3 % and $+3$ % and affects the estimates of irradiance and irradiation at ground level by the same relative quantity.

3.3 Extraterrestrial radiation received on a horizontal plane

Now that $E_{0N}(t)$ is known, the extraterrestrial radiation received on a horizontal plane located at the top of the atmosphere can be calculated. Let $E_0(t)$ and $H_0(t)$ be the extraterrestrial irradiance and irradiation, total or spectral, incident on this plane.

2 Meftah M., Dewitte S., Irbah A., Chevalier A., Conscience C., Crommelinck D., Janssen E., Mekaoui S., 2014. SOVAP/Picard, a spaceborne radiometer to measure the total solar irradiance. *Solar Physics*, 289, 1885–1899. doi:10.1007/s11207-013-0443-0.

3 The XXIXth International Astronomical Union General Assembly 2015. Resolution B3 on recommended nominal conversion constants for selected solar and planetary properties. Available at https://www.iau.org/static/resolutions/IAU2015_English.pdf, last access: 2020-06-04.

4 Kopp G., Lean J. L., 2011. A new, lower value of total solar irradiance: evidence and climate significance. *Geophysical Research Letters*, 38, L01706. doi:10.1029/2010GL045777.

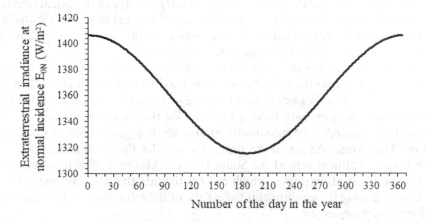

Figure 3.2 Annual profile of the extraterrestrial irradiance E_{0N} received by a plane always normal to the sun rays. The annual average is 1361 W m^{-2}.

If θ_S is the solar zenithal angle and Δt a period of time during which θ_S may be considered as constant, i.e., a period of 1 min or less, then:

$$E_0(t) = E_{0N}(t)\cos(\theta_S(t)) \tag{3.11}$$

$$H_0(t) = E_{0N}(t)\Delta t \cos(\theta_S(t)) \tag{3.12}$$

or at a given wavelength λ:

$$E_0(t,\lambda) = E_{0N}(t,\lambda)\cos(\theta_S(t)) \tag{3.13}$$

$$H_0(t,\lambda) = E_{0N}(t,\lambda)\Delta t \cos(\theta_S(t)) \tag{3.14}$$

These relationships show the importance of the solar zenithal angle θ_S on the extraterrestrial irradiance E_0 and irradiation H_0, and therefore on the irradiation and irradiation received on the ground. The greater θ_S, the more inclined on the plane the solar rays, and the smaller the irradiance, or irradiation, at the top of the atmosphere since $\cos(\theta_S)$ decreases with increasing θ_S.

At a given time, $E_{0N}(t)$, which is the irradiance received on the plane at normal incidence, does not depend on the latitude Φ, while $E_0(t)$ and $H_0(t)$ depend on it because the solar zenithal angle θ_S depends on Φ.

3.3.1 Typical values of hourly extraterrestrial total radiation

Extraterrestrial radiation received on a horizontal plane cannot be exceeded by radiation received on the ground, except in special cases. Knowing the typical values of extraterrestrial radiation makes it possible to quickly judge any value of solar radiation at ground level, which is often very practical. This is why I present to you, reader, some graphs to familiarize you with the typical values of extraterrestrial radiation.

Figure 3.3 exhibits the daily profile of the hourly average of the extraterrestrial total irradiance E_0 received on a horizontal surface at several latitudes for the 2 days of equinoxes and the 2 days of solstices. Also shown are the corresponding hourly irradiations H_0. The time in the day is the true solar time.

During equinoxes, the daily profile of E_0 is the same for the 2 days of equinox at given latitude. It is also the same between the two hemispheres at equivalent latitudes. This is why there is only one curve for these 2 days and that the curves for the northern hemisphere (in gray) are hidden by those for the southern hemisphere (in black) because they overlap. The maximum irradiance at a given hour is obtained at the equator. The closer you get to the poles, the smaller the irradiance; it is zero at the North Pole and almost zero at the South Pole on this curve. The irradiance is zero at the same hours at all latitudes, which reflects the fact that the daytime is the same at all latitudes during the equinoxes if the effects of the refraction of the solar rays by the atmosphere are neglected.

During the June solstice, the irradiance in the northern hemisphere is greater than that in the southern hemisphere at equivalent latitudes (central plot in Figure 3.3). It is the opposite during the December solstice (right plot). Irradiance at the poles is constant during the day, that at the South Pole during the December solstice being greater than that at the North Pole during the June solstice.

As during the equinoxes, the closer you get to the poles, the smaller the irradiance. It is a general trait independent of the day in the year, which is due to the influence of the solar zenithal angle θ_S on the received radiation. Similarly, you have noted the bell

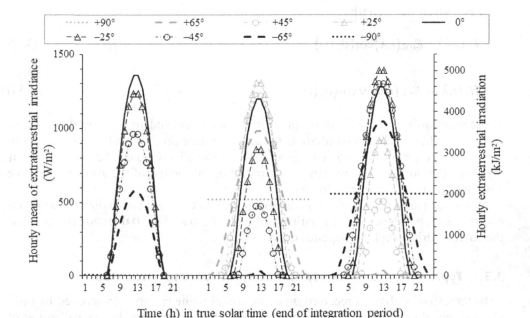

Figure 3.3 Daily profiles of the extraterrestrial total irradiance and irradiation received during 1 h on an horizontal plane, at several latitudes, during the days of equinoxes (left) and solstices of June (center) and December (right).

shape of the daily profile of the irradiance, except at the poles. Knowing that E_{0N} is almost constant during a day, this profile is due to the variation of θ_S during the day. These curves can be compared to the daily profiles of θ_S in Section 2.5, which have opposite shapes.

At equator, the maximum of hourly average irradiance is reached during equinoxes and is equal to 1358 W m^{-2}. During these days, the solar declination is equal to 0. In these conditions, the solar zenithal angle is:

$$\cos(\theta_S) = \cos(\omega) \tag{3.15}$$

At 12:00 TST, ω is equal to 0. Consequently, θ_S is zero, that is to say, that the radiation is at normal incidence (see Figure 2.8 or 2.5, Chapter 2) and therefore that the irradiance is maximum. The maximum reported in this curve is however slightly less than E_{TSI} (1361 W m^{-2}) because θ_S is zero only for, approximately, the minute centered on 12:00 TST, while the curve relates to hourly values, defined as follows:

$$E_0(t, \Delta t) = \int_{t-\Delta t}^{t} E_{0N}(u) \cos(\theta_S(u)) du \tag{3.16}$$

In other words, knowing that the daily profile of the irradiance exhibits a bell shape around 12:00 TST, the hourly average includes irradiances smaller than the peak reached at 12:00 TST and is therefore less than this maximum.

Reader, attentive as you are, it did not escape you that in Figure 3.3, the greatest hourly average of irradiance, equal to 1394 W m^{-2}, is obtained at 12:00 TST during the December solstice and at latitude $-25°$. This is due to the combination of the facts that the incidence is normal (see Figure 2.10 or 2.5) and that around this day, the sun–Earth distance is close to its minimum, and therefore, that E_{0N} is close to its maximum (see Figure 3.2).

Still by way of illustration, Figure 3.4 shows the variation of E_0 and H_0 when the sun is at its highest, that is to say, at 12:00 TST, at several latitudes as a function of the number of the day in the year. It therefore exhibits the variation of the maximum of the hourly average of irradiance during the year, since at 12:00 TST, the hour angle ω is zero and the solar zenithal angle θ_S is minimum. The accompanying table (Table 3.1) gives the annual averages as well as the minima and maxima of the irradiances at 12:00 TST. As plotted values are hourly averages, they are smaller than those that would be obtained over integration periods of 1 min and therefore less than E_{0N}.

The hourly average at 12:00 TST varies during the year in different ways according to latitudes. The closer the latitude to the poles, the less the irradiance at noon, as already noted in the previous figure. At the poles, the maximum irradiance is zero for half of the year. The maxima (Table 3.1) are reached around the summer solstice (June in the northern hemisphere and December in the southern one) and the minima around the winter solstice, except in the intertropical zone. In this zone, four extrema are observed instead of two. The maxima are located around the equinoxes and the minima around the solstices. When the latitude is close to the equator, the annual profile flattens, the winter minimum increases, and the summer maximum decreases to form a second minimum, while the values around the equinoxes increase and form two new maxima. This trend is observed on the curve at $-25°$. The irradiance at noon at the equator varies little during the year and offers the greatest annual average (1292 W m^{-2} in Table 3.1).

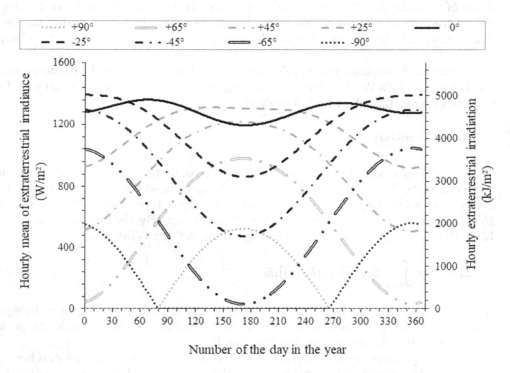

Figure 3.4 Annual profile of the extraterrestrial total irradiance and irradiation received during 1 h on a horizontal plane at 12:00 TST, at various latitudes.

The average is the same at equivalent latitudes in the northern and southern hemispheres (Table 3.1). The maxima are greater and the minima are smaller in the southern hemisphere than in the northern one. Hence, the differences between minimum and maximum are greater in the austral hemisphere than in the boreal one. The maximum irradiance is 1398 W m^{-2} and is reached around the December solstice at the Tropic of Capricorn (−23.45°). This maximum results from the combination of a normal incidence and the sun–Earth distance close to its minimum.

Table 3.1 Annual average, as well as minimum and maximum of the hourly average of the extraterrestrial total irradiance when the sun is at its highest during the day (12:00 TST), during a year, at various latitudes

Latitude	Irradiance (hourly mean, W m^{-2})			Latitude	Irradiance (hourly mean, W m^{-2})		
	Average	Minimum	Maximum		Average	Minimum	Maximum
90°	173	0	524	−90°	172	0	559
65°	546	32	981	−65°	546	30	1046
45°	914	507	1217	−45°	913	475	1298
25°	1171	920	1312	−25°	1171	862	1394
23.45°	1184	946	1313	−23.45°	1184	889	1398
11.5°	1265	1136	1343	−11.5°	1265	1068	1390
0°	1292	1196	1362				

3.3.2 Daily extraterrestrial radiation

The daily extraterrestrial radiation, whether total or spectral, does not depend on longitude but only on latitude and solar declination. It is easier to use the hour angle ω than time to demonstrate. Let H_{0day} be the daily extraterrestrial irradiation received on a horizontal plane. Let E_{0day} be the corresponding irradiance.

If $t_{sunrise}$, $\omega_{sunrise}$, t_{sunset}, and ω_{sunset} are respectively the time and the hour angle of the sunrise and sunset, H_{0day} is defined by:

$$H_{0day} = \int_{t_{sunrise}}^{t_{sunset}} E_{0N}(t)\cos(\theta_S(t))dt \tag{3.17}$$

Using equation (2.2) where the time is expressed in h, it comes:

$$H_{0day} = (12/\pi)\int_{\omega_{sunrise}}^{\omega_{sunset}} E_{0N}(\omega)\cos(\theta_S(\omega))d\omega \tag{3.18}$$

Since E_{0N} does not depend on ω during a day as a first approximation, it comes:

$$H_{0day} = (12/\pi)E_{0N}\int_{\omega_{sunrise}}^{\omega_{sunset}} \cos(\theta_S(\omega))d\omega \tag{3.19}$$

Using equations (2.6) and (2.19) and knowing that $\omega_{sunset}>0$ and $\omega_{sunset}=-\omega_{sunrise}$, it comes:

$$H_{0day} = (24/\pi)E_{0N}\left[\cos(\Phi)\cos(\delta)\sin(\omega_{sunset}) + \omega_{sunset}\sin(\Phi)\sin(\delta)\right] \tag{3.20}$$

or introducing the daytime S_0 (in h):

$$H_{0day} = E_{0N}\left[(24/\pi)\cos(\Phi)\cos(\delta)\sin(\omega_{sunset}) + S_0\sin(\Phi)\sin(\delta)\right] \tag{3.21}$$

It is thus demonstrated that H_{0day} depends only on latitude and solar declination.

Since time in equation (2.2) is expressed in h, H_{0day} is in Wh m^{-2} in equations (3.20) and (3.21). The daily irradiation is converted to J m^{-2} by multiplying the results of these equations by 3600 s. The daily average of irradiance E_{0day} is obtained in W m^{-2} by dividing the result of equation (3.20) or (3.21) by 24 h:

$$E_{0day} = E_{0N}\left[(1/\pi)\cos(\Phi)\cos(\delta)\sin(\omega_{sunset}) + (S_0/24)\sin(\Phi)\sin(\delta)\right] \tag{3.22}$$

Figure 3.5 exhibits the annual profile of E_{0day} at various latitudes. Table 3.2 gives the annual averages of the profiles presented in Figure 3.5 as well as the extrema of E_{0day}.

E_{0day} varies during the year in different ways according to latitudes. At equivalent latitudes, the curves in one hemisphere are opposite to those in the other hemisphere. Except the zone close to the equator, the annual profiles exhibit a bell shape with two extremes: a maximum reached around the summer solstice (June in the northern hemisphere and December in the southern one) and a minimum reached around the winter solstice. The variation around the mean value, i.e., the difference between maximum and minimum, decreases as the latitudes tend toward 0°. Table 3.2 shows that

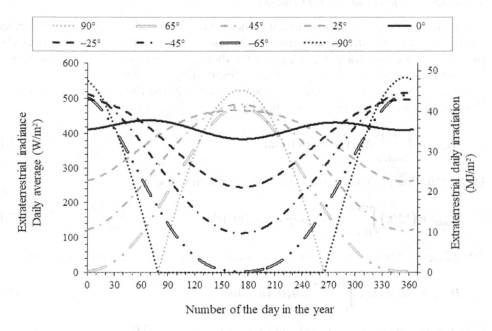

Figure 3.5 Annual profile of the extraterrestrial total irradiance and irradiation received on a horizontal plane during a day at several latitudes.

Table 3.2 Annual averages of the extraterrestrial total irradiance received by a horizontal plane at various latitudes, as well as minima and maxima of the daily mean

Latitude	Daily mean of irradiance (W m−2)			Latitude	Daily mean of irradiance (W m−2)		
	Mean	Minimum	Maximum		Mean	Minimum	Maximum
90°	172	0	524	−90°	172	0	559
65°	214	3	478	−65°	214	3	510
60°	236	24	476	−60°	236	23	509
45°	307	120	483	−45°	307	113	516
30°	365	227	475	−30°	365	213	506
25°	380	261	467	−25°	380	245	498
23.45°	384	271	463	−23.45°	384	255	495
11.5°	408	345	439	−11.5°	408	326	461
0°	416	384	438				

the annual average is the same at equivalent latitudes. For example, the average is the same at 45° and −45°.

The poles offer the greatest variation of E_{0day} during the year. E_{0day} is zero during about 6 months in the year. The maximum of E_{0day} is greater at the poles than at any other latitude. It is equal to 524 W m^{-2} at North Pole and 559 W m^{-2} at South Pole (Table 3.2). In contrast, the poles exhibit the lowest averages: 172 W m^{-2}.

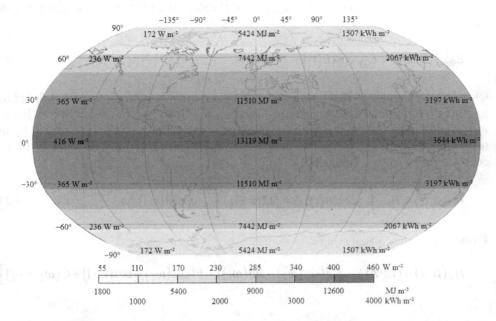

Figure 3.6 Map of the annual average of the extraterrestrial total irradiance received on a horizontal plane and the corresponding irradiation. Latitudes are shown on the left and longitudes at the top.

As the latitude tend toward 0°, the annual profile flattens, the winter minimum increases, the summer maximum decreases, and the values around the equinoxes increase. In the intertropical zone, the summer maximum decreases to form a second minimum, while the values around the equinoxes form two new maxima. The annual profile exhibits four extrema instead of two.

At equator, the solar zenithal angle at noon is small or even zero throughout the year. The daytime is always equal to 12 h (*cf.* Figure 2.7 in Chapter 2). It follows that E_{0day} varies little throughout the year (Figure 3.5) and that the equator offers the greatest annual average (416 W m^{-2} in Table 3.2). The variation of E_{0day} around the annual average is less than 13 % in relative value, with a minimum of 384 W m^{-2} and a maximum of 438 W m^{-2} (Table 3.2).

Figure 3.6 exhibits a map of the annual average of the extraterrestrial total irradiance received on a horizontal plane and the corresponding annual extraterrestrial total irradiation. It illustrates the latitudinal, or zonal, distribution; i.e., the values depend only on the latitude. The distribution is symmetrical with respect to the equator, as indicated in Table 3.2: The values are the same in both hemispheres at equivalent latitudes. The irradiance is the greatest at equator and decreases toward the poles.

3.4 Extraterrestrial radiation received on an inclined plane

The previous section deals with the extraterrestrial radiation received on a horizontal plane, total or spectral. This section deals with an inclined plane whose inclination is β and azimuth is α (*cf.* Figure 2.4, Chapter 2).

The extraterrestrial irradiance $E_{0incl}(\beta, \alpha, t)$ received on this inclined plane at time t is given by:

$$E_{0incl}(\beta,\alpha,t) = E_{0N} \cos(\theta(t)) \tag{3.23}$$

where θ is the angle of incidence of the sun rays, formed by the normal to the plane and the direction of the sun, given by equation (2.12). The extraterrestrial irradiation $H_0(\beta, \alpha)$ received during a period of time $[t_1, t_2]$, expressed in TST, is obtained by integrating equation (3.23), after a change of variable from t to the hour angle ω, on the corresponding interval $[\omega_1, \omega_2]$:

$$H_0(\beta,\alpha) = E_{0N} \int_{t_1}^{t_2} \cos(\theta(t)) dt = (12/\pi) E_{0N} \int_{\omega_1}^{\omega_2} \cos(\theta(\omega)) d\omega \tag{3.24}$$

Using equation (2.14), it comes:

$$H_0(\beta,\alpha) = (12/\pi) E_{0N} \left[A(\sin(\omega_2) - \sin(\omega_1)) - B(\cos(\omega_2) - \cos(\omega_1)) + C(\omega_2 - \omega_1) \right] \tag{3.25}$$

where the parameters A, B, and C are given by equation (2.15).

Given an opaque plane, the daily extraterrestrial irradiation $H_{0day}(\beta, \alpha)$ is given by:

$$H_{0day}(\beta,\alpha) = (12/\pi) E_{0N} \int_{\omega_1}^{\omega_2} \cos(\theta(\omega)) d\omega \tag{3.26}$$

where the integral limits ω_1 and ω_2 are solutions to $\cos(\theta) = 0$, equation that is resolved by means of equations (2.14) and (2.15).

Let take a few special cases. In the case of a vertical plane such as windows or walls of building oriented to the east, i.e., $\beta = \pi/2$ and $\alpha = \pi/2$, the limits ω_1 and ω_2 are respectively $\omega_{sunrise}$ and 0. The hour angle at sunrise $\omega_{sunrise}$ ($= -\omega_{sunset}$) is given by equation (2.19), and it comes:

$$A = 0; B = -\cos(\delta); C = 0$$

$$H_{0day}(\pi/2, \pi/2) = (12/\pi) E_{0N} \cos(\delta)(1 - \cos(\omega_{sunrise})) \tag{3.27}$$

The integral limits for a vertical plane oriented to the west ($\alpha = 3\pi/2$) are respectively 0 and ω_{sunset}. It follows that the plane oriented to the west receives the same daily irradiation than the plane oriented to the east:

$$H_{0day}(\pi/2, 3\pi/2) = H_{0day}(\pi/2, \pi/2) \tag{3.28}$$

In the case of a vertical plane oriented to the north or the south, the first step is the calculation of the hour angle ω_{E-W} corresponding to the time when the sun passes in the east–west plane containing the vertical plane. It happens for $\psi_S = \pi/2$, and the solution is obtained by equation (2.11) for $\tan(\psi_S)$:

$$\cos(\omega_{E-W}) = \tan(\delta)/\tan(\Phi) \quad \text{outside the poles} \tag{3.29}$$

if $\omega_{sunset} > \pi/2$, i.e., δ and Φ are of opposite signs, $\omega_{E-O} = \omega_{sunset}$

For a vertical plane oriented to the south ($\alpha = \pi$) and located in the northern hemisphere, the limits are $-\omega_{E-W}$ and ω_{E-W}. It comes:

$$A = \cos(\delta)\sin(\Phi) \; B = 0; \; C = \sin(\delta)\cos(\Phi)$$

$$H_{0day}(\pi/2, \pi) = (24/\pi) E_{0N} \left[\cos(\delta)\sin(\Phi)\sin(\omega_{E-W}) + \omega_{E-W} \sin(\delta)\cos(\Phi) \right] \tag{3.30}$$

For a plane located in the northern hemisphere oriented southward ($\alpha = \pi$) and whose inclination β is equal to the latitude Φ ($\beta = \Phi$), the limits are $-\omega_{E-W}$ and ω_{E-W}. It comes:

$$A = \cos(\delta); \; B = 0; \; C = 0$$

$$H_{0day}(\pi/2, \Phi) = (24/\pi) E_{0N} \cos(\delta)\sin(\omega_{E-W}) \tag{3.31}$$

Plants such as sunflowers and other radiation collectors can change orientation (tilt and azimuth) during the day. For example, concentration systems can rotate about an east–west, or north–south axis, with continuously adjusted positions in order to minimize the angle of incidence. There are also tracking systems fitted with two axes so that the plane is always normal to the direction of the sun. In these cases, the angle of incidence θ is given by:

$$\text{east} - \text{west axis: } \cos(\theta) = \left(1 - \cos(\delta)^2 \sin(\omega)^2\right)/2 \tag{3.32}$$

$$\text{north} - \text{south axis: } \cos(\theta) = \cos(\delta)$$

$$\text{two axes tracking: } \cos(\theta) = 1$$

Since time in equation (2.2) is expressed in h, $H_{0day}(\beta, \alpha)$ is expressed in Wh m^{-2} in the previous equations. The conversion to J m^{-2} is done by multiplying the result of the previous equations by 3600 s. The daily average of extraterrestrial irradiance $E_{0day}(\beta, \alpha)$ in W m^{-2} is obtained by dividing $H_{0day}(\beta, \alpha)$ resulting from one of the previous equations by 24 h.

Figure 3.7 exhibits the annual profile of the daily average of extraterrestrial irradiance $E_{0day}(\beta, \alpha)$ at latitude $+45°$ as a function of the number of the day in the year received by a plane (i) horizontal, (ii) inclined at 45° facing south, and (iii) vertical facing south. The daily average irradiance received on a vertical plane $E_{0day}(\beta = \pi/2, \alpha = \pi)$ is the same than that received on a horizontal plane $E_{0day}(\beta = 0, \alpha)$ during about half of the year including the boreal winter and is much less than during the other half of the year including the boreal summer. With an inclination of 45° ($\beta = \pi/4$), the plane facing south receives more radiation in annual average than the horizontal plane: 398 versus 308 W m^{-2}. The variation throughout the year is much smaller for the inclined plane than for the horizontal plane. The inclined plane exhibits a minimum of 348 W m^{-2} and two maxima of 439 W m^{-2}, while the horizontal plane exhibits a minimum of 121 W m^{-2} and a maximum of 485 W m^{-2}.

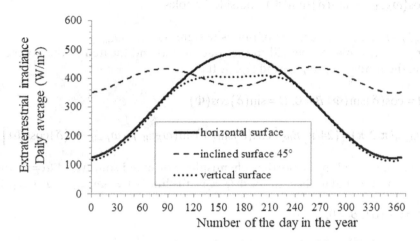

Figure 3.7 Annual profile of the daily mean of the extraterrestrial total irradiance at latitude 45° received by a plane (i) horizontal, (ii) inclined at 45° facing south, and (iii) vertical facing south.

3.5 Spectral distribution of the extraterrestrial radiation

Any object whose surface temperature is greater than 0 K emits electromagnetic radiation. Radiated energy is in the form of waves of different wavelengths. The distribution of the irradiance with the wavelength is called the spectral distribution of the irradiance. Its integral is total irradiance. Note that the usual notation for the wavelength is λ which should not be confused with the longitude.

Let $E_{0N}(\lambda)$ be the extraterrestrial spectral irradiance received on a plane at normal incidence at a given day at wavelength λ and $E_{TSI}(\lambda)$ be its annual average. It comes:

$$E_{0N} = \int_0^\infty E_{0N}(\lambda)d\lambda \tag{3.33}$$

$$E_{TSI} = \int_0^\infty E_{TSI}(\lambda)d\lambda \tag{3.34}$$

There is no dependence on the solar angles and the hour angle on the wavelength. If $E_0(\lambda)$ and $H_0(\lambda)$ denote respectively the spectral extraterrestrial irradiance and the spectral extraterrestrial irradiation received on a horizontal plane at wavelength λ, then:

$$E_0(\lambda) = E_{0N}(\lambda)\cos(\theta_S) \tag{3.35}$$

$$H_0(\lambda) = E_{0N}(\lambda)\Delta t \cos(\theta_S) \tag{3.36}$$

The previous equations are valid both for total irradiance E_0 or total irradiation H_0 and for their spectral counterpart $E_0(\lambda)$ or $H_0(\lambda)$.

The spectral irradiance is entirely determined by the emission properties and the temperature of the surface of the object, according to the laws of Kirchhoff and Planck. The radiation emitted by the sun is approximately that of a black body, that is to say, of a perfect emissive body, whose surface temperature is extremely high around

5780 K (about 5500 °C). Figure 3.8 exhibits a typical distribution of the extraterrestrial spectral irradiance $E_{0N}(\lambda)$. The spectral distribution indicates the amount of power at each wavelength. The area under the curve between two wavelengths λ_1 and λ_2 is the irradiance integrated over this interval.

The radiated power lies within a fairly limited range between 200 and 4000 nm, from X-rays to far infrared. This interval contains 99 % of the total solar irradiance TSI E_{TSI}, as it will be seen below. The spectral distribution is not uniform over this range and exhibits a marked peak around 500 nm. Half of the received power lies in the visible range, the rest being in the ultraviolet and in the near and middle infrared.

The distribution is not as smooth as that predicted for a perfect black body by Planck's law. There are several small troughs especially at wavelengths less than 800 nm. These troughs are called lines and are due to absorption and emission processes taking place in the sun and dependent on the wavelength. The sun does not emit as much energy in these absorption lines as in the neighboring wavelengths.

Figure 3.9 exhibits an excerpt of the spectral distribution, between 200 and 1500 nm for better readability of the lines; note their large number and their large amplitude at wavelengths between 200 and 400 nm, that is to say, in the ultraviolet (UV) range.

The spectral distribution $E_{0N}(\lambda)$ is not constant over time. I wrote previously that the inter-daily variations of E_{0N} can reach 5 W m^{-2}. These variations do not affect $E_{0N}(\lambda)$ equally. The visible and infrared domains are the least affected, while the ultraviolet is the most affected. The amplitude of the variations is 1–3 orders of magnitude greater in the UV range than in the visible or the infrared range. Knowledge and measurements of these variations in UV are still incomplete, although it is known that UV radiation is an important factor in the physical and chemical processes in the upper atmosphere.

Table 3.3 gives some integrated irradiances $E_{TSI}(\lambda_1, \lambda_2)$ over certain spectral intervals $[\lambda_1, \lambda_2]$. It reads that around 99 % of the total solar irradiance TSI E_{TSI} lies between 200 and 4000 nm. The UV-B irradiance in the interval [280, 315] nm is very small and is about 1 % of E_{TSI}. The irradiance in the UV-A interval [315, 400] nm is much greater and is about 6 % of E_{TSI}. About half of E_{TSI} lies in the visible part of the spectrum,

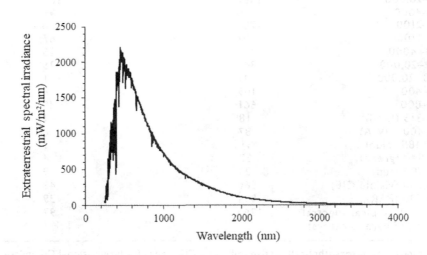

Figure 3.8 Typical distribution of the extraterrestrial spectral irradiance $E_{0N}(\lambda)$ received on a plane at normal incidence between 0 and 4000 nm.

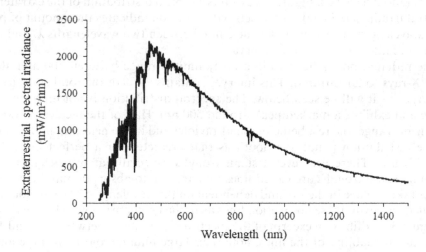

Figure 3.9 Typical distribution of the extraterrestrial spectral irradiance $E_{0N}(\lambda)$ received on a plane at normal incidence between 200 and 1500 nm.

i.e., in the range [380, 780] nm. The range of radiation for photosynthesis by green foliage of plants is [400, 700] nm; this radiation is called photosynthetically active radiation, abbreviated as PAR, and is on average equal to 39 % of E_{TSI}. If pyranometers were installed at the top of the atmosphere, they would measure approximately 93 % of E_{TSI}.

Table 3.3 Typical values of extraterrestrial irradiance received at normal incidence in various spectral intervals and their fraction of the total solar irradiance TSI

Spectral range (nm)	Irradiance in the interval (W m^{-2})	Fraction of E_{TSI} (%)
250–20,000	1361	100
250–4000	1350	99
380–2100	1202	88
400–1100	907	67
1100–4000	335	25
1100–20,000	346	25
4000–20,000	11	1
250–400	109	8
400–800	661	49
280–315 (UV-B)	18	1
315–400 (UV-A)	87	6
480–485 (blue)	11	1
510–540 (green)	57	4
620–700 (red)	123	9
380–780 (visible CIE)	660	48
400–700 (PAR)	534	39
330–2200 (typical spectral range of pyranometers)	1263	93

PAR stands for photosynthetically active radiation. CIE stands for International Commission on Illumination. Fractions are rounded up to the nearest integer.

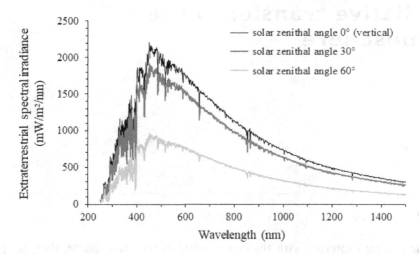

Figure 3.10 Typical distribution of the extraterrestrial spectral irradiance $E_0(\lambda)$ received on a horizontal plane between 200 and 1500 nm for three solar zenithal angles: 0°, 30°, and 60°.

The definition, or designation, of the spectral intervals can vary according to the field of application. For example, the visible range can denote the interval [380, 780], or [400, 700], or [400, 780], or [400, 800] nm in the scientific literature. The interval [380, 780] nm indicated here is that retained by the International Commission on Illumination – usually abbreviated as CIE for its French name, Commission Internationale de l'Eclairage. Another example is the use of the term broadband, which is not precisely defined and is often used to denote a wide spectral range. It is often used in the measurement of radiation by pyranometers that generally operate in the interval [300, 2500] nm approximately and whose outputs are called broadband measurements. This term can be used a little differently to denote a wide interval relative to a domain. For example, it can be read in scientific articles and other documents broadband UV that opposes spectral UV, or UV-B or UV-A.

As for E_0 and H_0 during a day, $E_0(\lambda)$ and $H_0(\lambda)$ reach their maximum when the solar zenithal angle θ_S is at its minimum, i.e., at 12:00 TST. $E_0(\lambda)$ and $H_0(\lambda)$ vary with θ_S and offer the same bell profile than E_0 and H_0 during the day. Figure 3.10 shows a typical spectral distribution of $E_0(\lambda)$ for three solar zenithal angles θ_S: 0°, 30°, and 60°, for wavelengths from 200 to 1500 nm. The influence of θ_S is clearly seen. The greater θ_S, the lower the spectral irradiance. The general shape is naturally the same for all angles since the distributions differ only by a factor $\cos(\theta_S)$ (equation 3.35) that does not depend on λ.

Chapter 4

Radiative transfer in the atmosphere

At a glance

The sun rays interact with the constituents of the atmosphere; they are partly absorbed and partly scattered. Radiation depletion is described using quantities such as absorption, scattering and attenuation coefficients, optical depth and thickness, transmittance, or visibility on the ground.

Only part of the extraterrestrial radiation reaches the ground, the rest being mainly absorbed or backscattered to the outer space by the constituents of the atmosphere. This proportion is around 70 %–80 % under clear-sky conditions, i.e., cloudless, and decreases in the presence of clouds. It also decreases as the concentration in the atmosphere of molecules, aerosols, water vapor, and other gases increases. The stratospheric ozone layer, between 20 and 40 km altitude, absorbs radiation at wavelengths below 280 nm. Air molecules strongly scatter short wavelengths up to 800 nm. The aerosols absorb a little the radiation and especially scatter it with a strong dependence on wavelength and type of aerosol.

Clouds mainly scatter radiation and are by far the main attenuators of solar radiation in its path to the ground. They exhibit very diverse geometric characteristics and optical properties. Optically thin clouds pass a significant fraction of the radiation downward, while optically thick clouds backscatter most of the radiation to space, resulting in a darkening of the sky for the observer at ground. The scattering by clouds is non-selective as a first approximation; i.e., it is the same whatever the wavelength.

During its downward path to the ground, the solar radiation interacts with the constituents of the atmosphere. This is called radiative transfer. The solar rays are partially absorbed, partially scattered, and possibly reflected by the ground, depending on the wavelength. The irradiance received on a plane at ground can represent only 70 %–80 % of that received at the top of the atmosphere under clear-sky conditions, i.e., under cloudless conditions. This fraction decreases if there are clouds. The rest of the radiation is backscattered to space.

Clouds play a key role in the extinction of radiation in its path to the ground. They are very diverse: Optically thin clouds pass a significant fraction of the radiation downward, while optically thick clouds obscure the sky by strongly scattering the solar rays toward space. Under cloudless sky conditions, called clear-sky conditions, the extinction of the radiation on its downward path to the ground is mainly due to absorption by gases, including water vapor, and scattering by molecules of the air and aerosols, which are particles suspended in the air such as soot or grains of sand.

This chapter deals with the radiative transfer in the atmosphere. The main optical phenomena responsible for the interactions between solar radiation and the atmosphere are absorption and scattering. I start with the presentation of the absorption of radiation by gases present in the atmosphere. I describe the phenomenon of scattering and how molecules in the air scatter the solar radiation. Then, I define attenuation, or extinction, and some associated quantities such as air mass, Linke turbidity factor, optical depth, or optical thickness. Next, the effects of air molecules and aerosols and the role of clouds, the latter being the main attenuators of solar radiation, are discussed for total and spectral radiation. Several diagrams illustrate the different paths that the sun rays can follow due to the scattering into the atmosphere. A summary of the contributions of the constituents of the atmosphere to the depletion of solar radiation is provided.

Before getting to the heart of the matter, allow me, reader, a brief description of the atmosphere. The latter is generally subdivided into six layers. The lowest layer is called the troposphere. It extends from the ground to an altitude of about 15 km at the equator and 8 km at the poles. The top of the troposphere is called the tropopause. In this layer, the density and the temperature of the air generally decrease as the altitude increases. The troposphere contains about 80 % of the total mass of the atmosphere. About half of this mass is contained between the ground and 6 km above sea level. It is the layer that has the most influence on solar radiation, in particular because it contains most of the clouds.

The stratosphere is above the tropopause up to an altitude of around 50 km. This 50 km limit is called stratopause. The third layer is the mesosphere, from 50 to 80 km, the summit of which is the mesopause. The other layers – the thermosphere, the ionosphere, and the exosphere – are only mentioned here for the record because they do not intervene, or in a negligible way, in the interactions between the atmosphere and the solar radiation. There is no well-defined border between space and atmosphere, as the atmosphere becomes less and less dense as the altitude increases. From the point of view of solar radiation, the mesopause can be considered as a limit of the atmosphere, that is to say, 80 km. In fact, this limit is of little importance to us. In practical terms, making radiative transfer calculations for the entire atmosphere from the ground up to 80, or 100 km, gives the same results.

What I call a clear sky in this book is a cloudless sky. It can be limpid if the atmosphere is not very loaded with water vapor and aerosols. If there is no aerosol and no water vapor – an ideal case – the atmosphere is called clean (without aerosols) and dry (low water vapor). Clear skies are called turbid if the aerosol load is significant.

4.1 Absorption by gases in the atmosphere

Absorption is a physical phenomenon present in the atmosphere. The energy absorbed by a constituent of the atmosphere, like gases, is converted into another form of energy and is no longer present in the radiation incident on the ground. Absorption can only occur at a specific wavelength or at a small range of neighboring wavelengths.

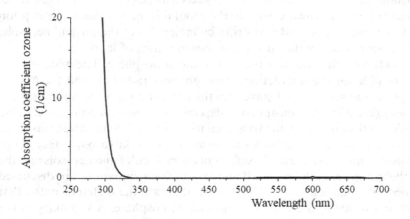

Figure 4.1 Typical spectral distribution of the absorption coefficient for ozone between 250 and 700 nm.

It is called selective absorption, and these wavelengths are called absorption lines. On the contrary, absorption can occur over a wide spectral range.

The absorption coefficient characterizes the intensity of absorption. The greater the coefficient, the more the radiation is absorbed. Its dimension is the inverse of a length. As an illustration, Figure 4.1 gives an example of the spectral distribution of the absorption coefficient for ozone between 250 and 700 nm. The absorption is very strong in the ultraviolet range at wavelengths less than 320 nm. It is much weaker at wavelengths around 600 nm and negligible elsewhere.

If F is the incident flux, dl the elementary distance in the absorbing medium, and a the absorption coefficient, the absorbed flux dF increases with the flux F and the coefficient:

$$dF = a\,F\,dl \tag{4.1}$$

The flux F' transmitted over a length l is:

$$F' = F \exp(-al) \tag{4.2}$$

The absorption coefficient can be defined for a wavelength, a range of wavelengths, or the entire spectrum, and is called spectral absorption coefficient, absorption coefficient for an interval, or total absorption coefficient, respectively.

The absorption of solar energy by particles and molecules leads to an increase in their own temperature. This increase in temperature induces an increase in the energy radiated by these absorbing bodies, which participates in terrestrial radiation. I remind you that terrestrial radiation is the energy sent back to space, also in the form of radiation but in longer wavelengths, between 3 and 80 μm, instead of the interval [0.2, 4] μm for solar radiation. It is through this process that the radiative balance is maintained between space, on the one hand, and the Earth including its atmosphere, on the other. Terrestrial radiation, also known as long waves, is emitted in all directions. It is itself partially absorbed and scattered by the atmosphere.

The ozone layer (O_3) is located in the stratosphere, between 20 and 40 km above sea level. It absorbs radiation in short wavelengths and therefore protects humans from the harmful effects of ultraviolet on their health. Ozone is also found near the ground, called tropospheric ozone, which results from photochemical interactions, i.e., under the influence of solar radiation, between nitrogen oxides and volatile organic compounds that are pollutants emitted mainly by human activities in transport and heating. This tropospheric ozone, which has a life span of a few days, is harmful to human health. It is generally overlooked in estimating the solar radiation received on the ground. However, it presents very important temporal and geographic variations that can possibly play a local role.

In the lower layers of the stratosphere and in the troposphere, other gases absorb solar radiation such as dioxygen (O_2), nitrogen oxides (NO_x), water vapor (H_2O), carbon dioxide (CO_2), and other gases, considered to be minor due to their low concentration such as methane (CH_4).

Figure 4.2 exhibits an example of the spectral distribution of the fraction of solar flux transmitted by the atmosphere, i.e., the ratio between the fluxes received on the ground and at the top of the atmosphere, from 250 to 4000 nm, for a moderately dry and clean atmosphere without aerosols. This fraction depends on the scattering by the molecules of the air at the shortest wavelengths less than 700 nm, as it will be seen later. But it mainly depends on the absorption by gases in the atmosphere. The graph exhibits more or less wide and deep indentations at certain wavelengths, in particular beyond about 1000 nm. They reflect the absorption by the gases that extract energy from the downwelling solar radiation in these wavelengths.

At wavelengths λ less than 280 nm, the solar radiation is absorbed by ozone (O_3) (see Figure 4.1) and the transmitted fraction is zero. Above 600 nm, there are numerous absorption lines, more or less wide, more or less deep, for dioxygen (O_2), water vapor (H_2O), carbon dioxide (CO_2), and methane (CH_4) for the main gases. For example, the

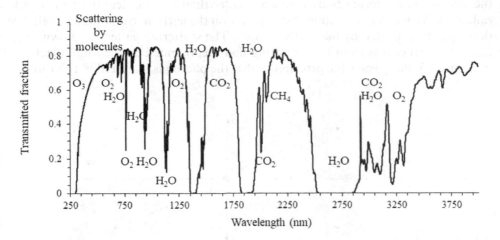

Figure 4.2 Example of spectral distribution of the fraction of the solar flux transmitted by a moderately dry and clean atmosphere, between 250 and 4000 nm. λ is the wavelength. O_3, ozone; O_2, dioxygen; H_2O, water vapor; CO_2, carbon dioxide; CH_4, methane.

transmitted fraction is zero around 1400 and around 1900 nm due to a strong absorption of radiation by water vapor.

4.2 Scattering – overview – case of the air molecules

Scattering is a physical phenomenon specific to the interactions between radiation and matter. It is present at all wavelengths, even if the intensity of the phenomenon can vary with the wavelength. Let be an incident radiation composed of photons. Any scattering body like a particle or a molecule located in the path of a photon deflects the latter, more or less strongly. If the photon takes a direction other than that of incidence, then the energy in the incident direction decreases. In this way, the scattering phenomenon subtracts energy from the incident radiation.

As in the case of absorption, a scattering coefficient b can be defined. If F is the incident flux, the flux F' transmitted over a length l in the scattering medium is:

$$F' = F \exp(-bl) \tag{4.3}$$

The dimension of this coefficient b is the inverse of a length. b can be defined for a wavelength, a range of wavelengths, or the entire spectrum; it is called spectral scattering coefficient, scattering coefficient for an interval, or total scattering coefficient, respectively.

The way in which a body scatters energy according to the direction is called the scattering pattern or the phase function. The phase function gives the probability that a photon is scattered in a given direction. If the scattering is isotropic, the energy is scattered equally in all directions. The phase function and the scattering pattern depend on the body itself, its nature, its shape, its size, and its surface properties as well as on the wavelength.

Figure 4.3 is a schematic view of a possible scattering pattern in the case where the size of the scattering body is much smaller than the wavelength of the incident radiation. An incident radiation (black arrow on the left) can be scattered in all directions, partly indicated by the dotted arrows. The scattering pattern is shown in gray line. The further it is from the scattering body represented by the small black circle in Figure 4.3, the greater the probability that the photon is scattered in this direction.

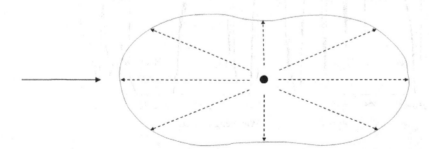

Figure 4.3 Schematic view of a possible scattering pattern in the case where the scattering body (small black circle) is much smaller than the wavelength of the incident radiation shown by the solid arrow.

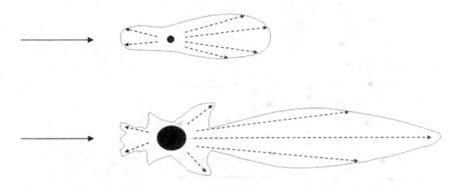

Figure 4.4 Schematic view of possible scattering patterns in the case where the size of the scattering body (small black circle) is comparable to or slightly greater than the wavelength of the incident radiation shown by the solid arrow.

In this diagram, the scattering pattern is almost an ellipse: The probability that the photon is scattered forward, i.e., in the incident direction, or backward, is greater than the probability of scattering at 90° or −90°, i.e., perpendicular to the incident direction. This pattern is typical of scattering patterns for gas molecules in the atmosphere.

Two other scattering patterns are given in Figure 4.4 for particles larger than the molecule in the previous case. In both diagrams, the scattering patterns are very elongated forward, which means that the probability is very high for the photons to continue in the incident direction. In the upper diagram, photons are unlikely to be scattered in perpendicular directions; more often they will be scattered mainly forward or backward. As the size of the particle increases, the probability that the photons are scattered forward generally increases, with scattering patterns that may have complex shapes as shown in the lower diagram. The scattering patterns shown here are only schematic representations. In reality, the particles are not smooth spheres and the scattering patterns are more complicated and asymmetrical.

The atmosphere contains many molecules and particles. Frequently, one of these scattering bodies scatters photons that have already been scattered by other bodies, and so on. This multiple scattering is illustrated in Figure 4.5. The two lines frame the incident radiation beam, formed by parallel rays. Photons are scattered several times in this beam. They can remain there or leave it, or even reintegrate it after multiple scattering outside the beam.

Scattering by bodies of comparable size or slightly greater than the wavelength of the incident radiation can be described by Mie's law.[1] The latter describes the scattering by a homogeneous sphere. Particles in the atmosphere are not homogeneous perfect spheres, and Mie's law does not apply perfectly. However, it offers a first approximation of the scattering by aerosols, these particles, solid or liquid, suspended in the atmosphere, often located in the lower layers of the atmosphere, from the ground up to about 2 km altitude.

1 Gustav Mie was a German physicist (1868–1957).

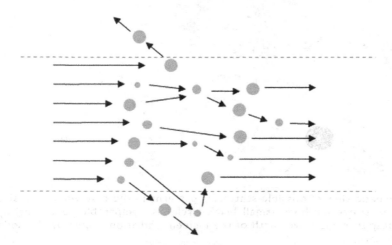

Figure 4.5 Illustration of multiple scattering within an atmospheric column.

When the scattering bodies are much smaller than the incident wavelength, scattering can be described by Rayleigh's law.[2] This is the case for air molecules and other gases; this is why a clean and dry atmosphere that contains only such molecules is sometimes called the Rayleigh atmosphere. According to this law, the variation of the scattering coefficient is in λ^{-4}, where λ is the wavelength. This means that the smaller the wavelength, the more important the scattering.

Scattering by molecules is well marked at wavelengths less than about 700 nm. The influence of this scattering can be seen in Figure 4.2 showing an example of the spectral distribution of the fraction of solar flux transmitted by the atmosphere. This results in a rounded shape of the distribution up to 700 nm. Incident radiation loses more energy in its downwelling path to the ground at short wavelengths than at long ones.

In the visible part of the solar spectrum, blue photons are more scattered than those with longer wavelengths, the least scattered being red photons. As shown in Figure 4.3, the scattering pattern for air molecules is almost an ellipse; i.e., the probability of scattering is noticeable for any direction, even if forward or backward scattering exhibits the highest probabilities. As a result, blue photons are scattered in all directions, much more than longer photons. They form an important part of the radiances coming from all directions of the sky vault, except the direction of the sun. When the sky is very clear, the portion of the sky far from the direction of the sun appears dark blue to the observer because the radiances mainly include blue photons. As water vapor increases, absorption and scattering increase at all wavelengths that are more multiscattered than in the case of a dry atmosphere. The sky vault then appears rather white.

---

2 John William Strutt Rayleigh (1842–1919) was a British physicist known for his brilliant works in various domains.

4.3 Attenuation – extinction

In this section, I present the attenuation or extinction as well as several quantities associated with this notion of attenuation and often used in the fields of radiative transfer or solar radiation. All of these quantities can be defined for a wavelength, a range of wavelengths, or the entire spectrum.

The two phenomena – absorption and scattering – remove energy from the incident solar radiation in its downwelling path to the ground. This decrease in radiation is called attenuation, or extinction. Like absorption and scattering, attenuation depends on the wavelength. An attenuation coefficient, c, also called extinction coefficient is defined, such that if F is the incident flux, the flux F' transmitted over a length l in an absorbing–scattering medium is:

$$F' = F \exp(-cl) \tag{4.4}$$

with $c = a + b$.

The dimension of c is the inverse of a length.

4.4 Quantities related to the attenuation

4.4.1 Air mass

The extinction of a radiation beam depends on the number and sizes of particles and molecules on the path, on the volumetric density of the air, and on the path length across the attenuating medium. The relative air mass characterizes this extinction for any incidence relative to that at zenith, i.e., for a vertical path. It is often abbreviated as air mass and noted m. It is dimensionless. This term is not to be confused to the air mass used in meteorology. The air mass can be defined for a given atmospheric layer, or for a given constituent, e.g., the ozone air mass, or for the entire atmospheric column from sea level to the top. By default, the latter prevails.

At normal incidence, i.e., when the solar zenithal angle θ_S is zero, the air mass is equal to 1 by definition. The closer the sun to the horizon, the longer the path through the atmosphere, and the more important the extinction of the beam radiation. When θ_S is 60°, or the solar elevation angle is 30°, the air mass is 2. It reaches approximately 40 when the sun is on the horizon. The simplest equation of the air mass is:

$$m = 1/\cos(\theta_S) \tag{4.5}$$

The results are quite acceptable in most cases when θ_S is less than 70° in the case of total radiation. More complex formulations have been proposed, some of them taking into account changes in density in the atmospheric column and depending on the spectral range under concern. For the total radiation, the Kasten and Young[3] formula gives accurate enough results even when the sun is on the horizon:

3 Kasten F., Young A. T., 1989. Revised optical air mass tables and approximation formula. *Applied Optics*, 28, 4735–4738. doi:10.1364/AO.28.004735.

$$m = 1 \Big/ \left[\cos(\theta_S) + 0.50572(6.07995° + 90° - \theta_S)^{-1.634} \right] \tag{4.6}$$

where θ_S is in degrees.

The geometrical thickness of the atmosphere depends on the elevation, or altitude, of the site above the mean sea level. The air mass is given for the mean sea level, i.e., for an atmospheric pressure at surface P_0 equal to 101.324 hPa. At a given site, the atmospheric pressure at surface P is related to the elevation of the site. Since the elevation and the air pressure are correlated, a usual correction of the air mass for difference in elevation is given by:

$$m_{corrected} = mP/P_0 \tag{4.7}$$

4.4.2 Optical depth – optical thickness

The optical depth τ is a measure of the attenuation of the radiation during its vertical path though the atmosphere, i.e., when θ_S is zero. It is the natural logarithm (ln) of the ratio of the incident flux $F_{incident}$ at the top of an atmosphere layer to the flux $F_{transmitted}$ transmitted by this layer:

$$\tau = \ln\left(F_{incident} / F_{transmitted} \right) \tag{4.8}$$

The atmosphere layer can be composed of molecules of air, aerosols, or clouds. The greater the attenuation in this layer, the greater the optical depth. It is positive and dimensionless. The optical depth of aerosols typically varies between 0 and 5 as it will be seen later. The range of variation of the optical depth of clouds is much larger, from 0 to more than 100. An observer on the ground can hardly see the sun when the optical depth of the cloud is around 3.

Using the previous equations, a relationship is obtained between the optical depth τ and the attenuation coefficient c on a vertical path:

$$\tau = \exp(-c) \tag{4.9}$$

or

$$\tau = \exp\left(-(a+b)\right) = \tau_{absorption} + \tau_{scattering} \tag{4.10}$$

where $\tau_{absorption}$ and $\tau_{scattering}$ are the optical depths due respectively to absorption and scattering. In other words, the optical depths add up along the radiation beam. In the case of an atmosphere column containing absorbing gases, scattering air molecules, and scattering aerosols, Beer–Lambert's law links the incident flux $F_{incident}$ and the transmitted flux $F_{transmitted}$:

$$F_{transmitted} = F_{incident} \exp\left(-m\left(\tau_{gases} + \tau_{molecules} + \tau_{aerosols}\right)\right) \tag{4.11}$$

where τ_{gases}, $\tau_{molecules}$, and $\tau_{aerosols}$ are respectively the optical depths of the absorbing gases, air molecules, and aerosols. Each optical depth can itself be broken down into the sum of optical depths. For example, the optical depth of gases is the sum of that for ozone, that for dioxygen, etc.

Some authors use the terms optical depth and optical thickness interchangeably. This is not the general case. For example, the World Meteorological Organization limits the use of optical thickness to oblique paths.[4] If τ_{depth} and $\tau_{thickness}$ denote respectively the optical depth and thickness, the following relationship holds:

$$\tau_{thickness} = \tau_{depth}/\cos(\theta_S) \tag{4.12}$$

4.4.3 Transmittance

The transmittance T quantifies the amount of energy transmitted by an atmospheric column compared to that incident at the top of this column. Let $F_{incident}$ and $F_{transmitted}$ be respectively the incident flux at the top of the atmosphere and the flux received on the ground. The transmittance T is defined as the ratio of these two fluxes:

$$T = F_{transmitted}/F_{incident} \tag{4.13}$$

$$\text{or}\quad T = \exp(-\tau m) \tag{4.14}$$

where τ is the optical depth (vertical path) of the atmosphere column and m is the air mass. The transmittance is dimensionless. It varies between 0 (totally opaque medium) and 1 (fully transparent medium). It is the transmittance that is plotted in Figure 4.2, which I then called transmitted fraction, since transmittance was not defined yet.

Since the optical depths add up along the radiation beam, the transmittances multiply. Given an atmosphere column containing absorbing gases, scattering air molecules, and scattering aerosols, the transmittances T_{gases}, $T_{molecules}$, and $T_{aerosols}$ may be defined for absorbing gases, scattering molecules, and scattering aerosols, respectively:

$$F_{transmitted} = F_{incident} T_{gases} T_{molecules} T_{aerosols} \tag{4.15}$$

Each transmittance can itself be broken down into the product of transmittances. For example, the transmittance of gases is the product of that of ozone, that of dioxygen, etc.

I draw your attention to the fact that transmittance is linked to the optical depth and to the air mass by an exponential. A small variation in one or other of the variables results in a greater variation in transmittance. Let me take a few examples. Assume a large solar zenithal angle θ_S equal to 75°, and let me use the approximation $m = 1/\cos(\theta_S)$. If the optical depth is 0.1, which is very small, then the transmittance is 0.68. If the optical depth is slightly greater and equal to 0.5, the transmittance is 0.14, i.e., much smaller. In another example, the optical depth is now set to 1. When θ_S is respectively 30° and 75°, T is 0.32 and 0.02. These few examples demonstrate the importance of the optical depth and the solar zenithal angle on the transmittance.

---

4 Guide to Instruments and Methods of Observation. WMO-No. 8, 2018 edition, 2018, World Meteorological Organization, Geneva, Switzerland. [Vol 1 Measurement of Meteorological Variables, Chapter 7, Measurement of Radiation.]

4.4.4 Linke turbidity factor

The Linke turbidity factor[5] synthesizes the atmospheric attenuation under cloudless sky conditions. It is often noted T_L and is dimensionless. It takes into account absorption and scattering by water vapor and aerosols relative to that of a clean and dry atmosphere. T_L is 1 for this atmosphere. Given an air mass, the Linke turbidity factor is the number of clean and dry atmospheres that should theoretically be superimposed to obtain an attenuation equivalent to that observed for the atmosphere considered. The greater the attenuation by the atmosphere, the greater T_L.

Unlike the attenuation, the air mass, the transmittance, or the optical depth, which can be defined for a wavelength, a range of wavelengths, or the entire spectrum, the Linke turbidity factor applies to the total radiation though it could be spectrally defined in theory.

The Linke turbidity factor is often given for an air mass equal to 2, i.e., $\theta_S = 60°$. A value of 3 is typical for Africa and Europe. The Linke turbidity factor can reach 7 or more, in the case of polluted areas, such as urban areas with heavy road traffic. It is a very convenient quantity for synthesizing the attenuation of the cloudless atmosphere and is often used by engineers and other practitioners in various fields of solar radiation.

If τ_{mixed_gases} is the optical depth due to the absorption by mixed gases (mainly CO_2, O_2) in the atmosphere, τ_{ozone} that due to ozone, and $\tau_{scattering\_molecules}$ that due to the scattering by the air molecules, the optical depth of a clean and dry atmosphere τ_{clean_dry} is given by:

$$\tau_{clean\_dry} = \tau_{mixed\_gases} + \tau_{ozone} + \tau_{scattering\_molecules} \tag{4.16}$$

Let τ_{water_vapor} be the optical depth due to the absorption by water vapor and $\tau_{aerosols}$ that due to the scattering by aerosols. The optical depth τ of this atmosphere is then:

$$\tau = \tau_{clean\_dry} + \tau_{water\_vapour} + \tau_{aerosols} \tag{4.17}$$

and it comes:

$$T_L = \tau / \tau_{clean\_dry} \tag{4.18}$$

For illustration purposes, Figure 4.6 exhibits typical monthly values of the Linke turbidity factor at several locations. The site of Barrow in Alaska in the United States (latitude: 71.32°; longitude: −156.61°; elevation: 10 m) is north of the northern polar circle. It experiences very clear skies in boreal winter, and T_L is small and close to 2 from November to January. T_L increases up to around 4 in summer as the content in water vapor increases in the atmosphere. Bondville (latitude: 40.11°; longitude: −88.37°; elevation: 220 m) is a small city in rural landscape in Illinois in the United States that experiences a temperate climate. Similarly to Barrow, T_L reaches its minimum in

---

5 Franz (exactly Karl Wilhelm Franz) Linke was a German geophysicist and meteorologist (1878–1944). He defined the turbidity factor (Trünbungsfaktor) in 1932. As Linke is a German name, the final e must be pronounced.

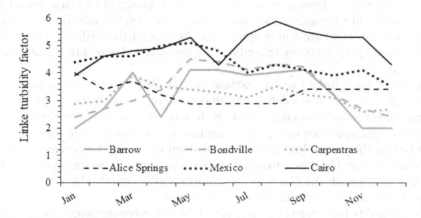

Figure 4.6 Typical monthly values of the Linke turbidity factor at several locations. (Source: SoDa Service (www.soda-pro.com).)

winter ($T_L \approx 2.5$) and its maximum ($T_L \approx 4.5$) in summer. Carpentras (latitude: 44.06°; longitude: 5.05°; elevation: 100 m) is also a small city in rural landscape with a temperate climate in Provence, which is the sunniest part of France. Like the previous two sites, the minimum is reached in winter and the maximum in summer, but unlike the other two, its variation throughout the year is small. T_L ranges between 2.6 and 3.9. Actually, Provence is known for its often very clear sky. Alice Springs (latitude: −23.80°; longitude: 133.89°; elevation: 550 m) is a small city situated roughly in the center of Australia in an arid environment. It experiences a subtropical hot desert climate. Expectedly, the skies are often clear all year round, and T_L ranges from 2.9 to 4. The minimum is reached in austral winter (June–August) and the maximum is observed in summer when the atmospheric content in water vapor increases and dust storms occur.

The two other sites are megacities and are air-polluted areas. T_L at Mexico City, the capital of Mexico (latitude: 19.43°; longitude: −99.13°; elevation greater than 2200 m), ranges between 4 and 5. A greater T_L could be expected due to air pollution, but it is partly offset by the high altitude of the city. Cairo, the capital of Egypt (latitude: 30.05°; longitude: 31.24°; elevation: 20 m), has almost 20 million inhabitants and millions of vehicles and thousands of factories. It has an arid climate with increased air moist in summer. T_L is about 4 in January and increases till May, partly because of the occurrence of dust storms. After a fall in June, it increases again up to 6 in August, partly because of the increase in air moist, and partly because at the end of the growing season, farmers used to burn rice straw, thus adding to the air pollution in summer and thus increasing the depletion of the solar radiation reaching the ground. The values given here for illustration are indicative. They may evolve with changing urban policies and other constraints.

4.4.5 Visibility

Unlike the Linke turbidity factor, visibility is not just about cloudless skies. It quantifies the horizontal attenuation of the atmosphere, cloudy or not, at ground level.

Visibility is the greatest horizontal distance at which an observer can distinguish a black object of sufficient size against the background of the sky. It is usually expressed in km. The greater the attenuation at ground level, the smaller the visibility. Given its definition, the visibility is a quantity related to the human visual perception and is not spectrally defined.

Visibility is very much linked to the particle content in the atmosphere but also to the presence of fogs, which are defined as clouds in contact with the ground. For example, it can be very small, less than 1 km, in the case of fogs. The combination of suspended desert dust and high water vapor content in the coastal area means that visibility can be less than a few km around the Arabian Gulf, also known as the Persian Gulf or Gulf of Iran. On the contrary, from Sophia Antipolis, in the southeast of France, you can distinguish the Corsican Mountains 200 km apart when the atmosphere is clear, particularly early in the morning in winter.

Visibility is a quantity measured at all airports. It is very often reported in meteorological bulletins and related databases. It is a convenient way to get a first idea of how the atmosphere attenuates the radiation in a given place and time. Certain numerical models modeling the radiative transfer in the atmosphere allow the visibility as input to quantify the extinction of the atmosphere without cloud. For a beginner, it is indeed an easier quantity to grasp than the optical depth of aerosols at 500 nm, or another wavelength, or the water vapor content of the atmospheric column. For example, a visibility of 50 km may be input to the model for a clear sky, or 10 km for a turbid sky loaded with aerosols and water vapor.

4.5 Effects of aerosols on radiation

Aerosols are particles, solid or liquid, suspended in the atmosphere. Their origins are natural and anthropic. Among the aerosols of natural origin, are particles of sea salt coming from the sea spray, particles of sand torn from the neighboring deserts by dust storms, and chemical compounds emitted by the biosphere, the last term designating all living organisms and their living environments. Aerosols of anthropogenic origin come mainly from industrial and agricultural activities, transport, and the burning of biomass, producing sulfates, nitrates, and carbon aerosols, like soot and others of organic origin.

The effects of aerosols alone on the solar radiation are often grouped under the term turbidity. If the aerosol load in the cloudless atmosphere is low, the attenuation is low and the sky is said to be very clear or limpid. When the aerosol load is high, the sky is said to be turbid. Aerosols play a very important role in the scattering and, to a lesser extent, the absorption of solar radiation under clear-sky conditions. The absorption and scattering capacities are not the same for all aerosols. For example, sulfates are rather scattering and soot rather absorbing.

Aerosols are very often located in the lower layers of the atmosphere, from the ground to about 2 km above sea level. Roughly speaking, the higher the altitude, the less aerosols, and the greater the transmittance. However, this is not a general rule. Particularities of the sites in altitude must be taken into account. As an illustration, I have extracted time series of the aerosol optical depth at 550 nm, every 3 h, during the year 2019 from the Copernicus Atmosphere Monitoring Service (CAMS) at two close locations in Peru (Figure 4.7). The city of Cuzco (latitude: −13.52°; longitude: −71.98°)

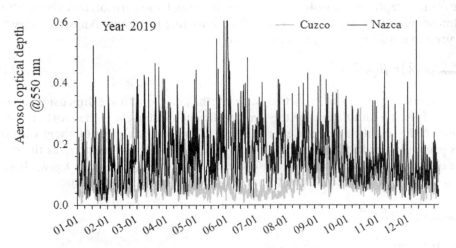

Figure 4.7 Aerosol optical depth at 550 nm every 3 h during the year 2019 at Cuzco and Nazca, estimated by the Copernicus Atmosphere Monitoring Service (CAMS). (Source: SoDa Service (www.soda-pro.com).)

is in the Andes Cordillera at an elevation of 3300 m above the mean sea level, while the city of Nazca is on the foothills close to the seashore (latitude: −14.83°; longitude: −74.94°) and has an elevation of 500 m. Thus, one would expect a lower aerosol optical depth in Cuzco than in Nazca most of the time. It is not entirely the case in Figure 4.7: The gray curve for Cuzco is often close to the black curve for Nazca. As a whole, the aerosol optical depth in Cuzco tends to be less than in Nazca. This is especially visible in austral winter, from June to August. Other causes than altitude intervene, and there are many peaks in aerosol optical depth in Cuzco. This is due to the fact that Cuzco is a city of approximately 400,000 inhabitants with agricultural activities, while Nazca has ten times less inhabitants.

Volcanic explosions can introduce ash at several tens of kilometers above sea level, which will remain suspended for several days or even months in the upper layers. The exceptional eruption of the Pinatubo volcano in the Philippines in 1991 released a considerable amount of volcanic ash into the atmosphere, which affected the radiation balance and therefore the climate, for 2–3 years.

Atmospheric aerosols show significant variations in chemical composition, volume concentration, shape, and size. They therefore offer significant variations in absorption and scattering. As seen previously, the scattering depends on the size of the scattering particles and is a function of the wavelength λ. The aerosols have very variable sizes, ranging between approximately 0.1 and 10 μm. Sulfates, resulting from pollution, and carbonaceous materials, resulting from combustion, are generally small, while sea salts and grains of sand are larger. But even for particles of the same type, there is a large variation in size within a sample. In addition, the humidification of the particles increases their size, which will vary according to the local humidity at the altitude concerned. Currently, the effects of aerosols on solar radiation are difficult to estimate and to model.

The optical depth of aerosols $\tau_{aerosols}$ varies between 0 (no aerosol) and about 5 (turbid atmosphere). It depends on λ. At the time of writing this book, Ångström's law[6] is often used to model $\tau_{aerosols}(\lambda)$ as a function of λ:

$$\tau_{aerosols}(\lambda) = \beta(\lambda/\lambda_1)^{-\alpha} \tag{4.19}$$

In this equation, λ_1 and λ are any two wavelengths and α and β are dimensionless coefficients. I used α and β here for the two coefficients because it is the most common notation in the scientific literature for this law, at the risk of confusing them with the angles generally used for the inclined plane. The coefficient β characterizes the magnitude of the optical depth of the aerosols and is equal to $\tau_{aerosols}$ when $\lambda = \lambda_1$. If λ_1 is equal to 1 µm, then β is called Ångström turbidity coefficient, and it comes:

$$\tau_{aerosols}(\lambda) = \beta\lambda^{-\alpha} \tag{4.20}$$

where λ is in µm.

The coefficient α is called Ångström exponent. It describes how the optical depth $\tau_{aerosols}$ depends on λ. It depends on the sizes of the particles and their nature: In general, the smaller the particle, the greater α. Large variations in α are observed in nature, with possible negative values. α typically varies between -1 and 4. When the Ångström exponent is zero, the optical depth $\tau_{aerosols}$ does not depend on λ, according to equation (4.19). If α is positive, the optical depth decreases with λ. Conversely, if α is negative, the optical depth increases with λ.

In general, $\tau_{aerosols}$ is often known at a single wavelength, typically 500 or 550 nm. $\tau_{aerosols}(\lambda)$ is then calculated at any wavelength with equation (4.19) assuming α is known. If $\tau_{aerosols}$ is known at two wavelengths λ_1 and λ_2, α can be calculated using the following equation:

$$\alpha = -\log\left[\tau_{aerosols}(\lambda_1)/\tau_{aerosols}(\lambda_2)\right]/\log(\lambda_1/\lambda_2) \tag{4.21}$$

and then $\tau_{aerosols}(\lambda)$ can be calculated at any wavelength.

For illustration purposes and as for Peruvian sites, I have extracted time series of the aerosol optical depths, every 3 h, during the year 2019 from the CAMS at several locations. Figure 4.8 exhibits the aerosol optical depth at 550 nm at two polluted megacities: Beijing, the capital of China (latitude: 39.91°; longitude: -116.36°; elevation: 40 m); and Mexico City (Mexico, latitude: 19.43°; longitude: -99.13°; elevation greater than 2200 m). A striking feature of these graphs is the large variation in $\tau_{aerosols}$ within a day and from day to day. This induces even greater variations in transmittance. At Beijing, the maximum in $\tau_{aerosols}$ is 5.0. $\tau_{aerosols}$ is often greater than 0.4; i.e., the corresponding transmittance at 550 nm due to aerosols only is often less than 0.6 when the

---

6 Ångström! What a dynasty of Swedish physicists, astronomers, geophysicists, and meteorologists! The grandfather, Anders Jonas (1814–1874), studied solar radiation and absorption lines (the so-called Fraunhofer lines). He is well known for having established a unit of length, the Ångström that is 10^{-10} m. The father, Knut Johan (1857–1910), also studied the solar radiation. It is to Anders Knutsson (1888–1981) that we owe this very convenient law on optical depths of aerosols. He also proposed a relationship between the solar radiation and the sunshine duration and invented the pyranometer.

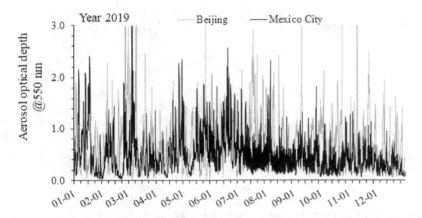

Figure 4.8 Aerosol optical depth at 550 nm every 3 h during the year 2019 at Beijing and Mexico City estimated by the CAMS. (Source: SoDa Service (www.soda-pro.com).)

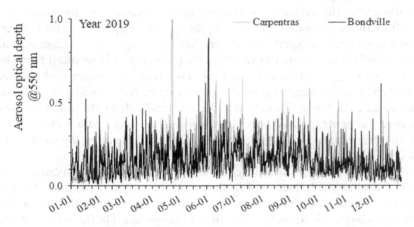

Figure 4.9 Aerosol optical depth at 550 nm every 3 h during the year 2019 at Carpentras and Bondville, estimated by the CAMS. (Source: SoDa Service (www.soda-pro.com).)

solar zenithal angle θ_S is 30°. The maximum of $\tau_{aerosols}$ at Mexico City is 3.0. Overall, $\tau_{aerosols}$ at Mexico City is less than $\tau_{aerosols}$ at Beijing. This could be due in part to the very high altitude of Mexico City, above 2000 m.

In contrast to the previous case, I drew in Figure 4.9 the same graphs but for two unpolluted sites in rural areas: Carpentras (France, latitude: 44.06°; longitude: 5.05°; elevation: 100 m) and Bondville (United States, latitude: 40.11°; longitude: −88.37°; elevation: 220 m). There are also great variations in $\tau_{aerosols}$ within a day and from day to day. However, the range of values taken by $\tau_{aerosols}$ is much smaller than in the previous case of Beijing and Mexico City. The maximum of $\tau_{aerosols}$ is 1.4 in Carpentras and 0.9 in Bondville. At both sites, $\tau_{aerosols}$ is often in the range [0.1, 0.2]. When θ_S is 30°, the

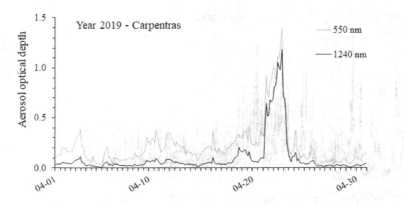

Figure 4.10 Aerosol optical depths at 550 and 1240 nm every 3 h at Carpentras in April 2019 estimated by the CAMS. (Source: SoDa Service (www.soda-pro. com).)

corresponding transmittance at 550 nm due to aerosols only is 0.9 and 0.8 for $\tau_{aerosols}$ equal to 0.1 and 0.2, respectively.

Figure 4.10 illustrates the influence of the size of the aerosols on the variation in $\tau_{aerosols}$ with λ. It exhibits the aerosol optical depths τ_{550} and τ_{1240} at 550 and 1240 nm, respectively, at Carpentras in April 2019. Note that $\tau_{550} > \tau_{1240}$. In this case and according to Ångström's law, α is greater than 1, which corresponds to small particles, and the variation in $\tau_{aerosols}$ with λ is important. This is in agreement with the fact that overall, pollution in Carpentras is of anthropic origin. Nevertheless, τ_{1240} is very close to τ_{550} around 22–23 April. During these days, a wind laden with sand blew from North Africa over Carpentras. Sand particles exhibit large sizes and α is close to 1 in accordance with observations and Ångström's law. During these days, the variation in $\tau_{aerosols}$ with λ was weak.

Avid reader, allow me to digress on the complex relationships between climate and aerosols and to underline the influence of the latter on the radiative balance. Aerosols scatter the photons back to space, thus increasing the planetary albedo that will be detailed in the next chapter, leading to a cooling of the surface. On the contrary, they also absorb the solar radiation, especially soot, and contribute to a warming of the atmosphere and hence to a possible warming of the surface. There are also indirect effects of aerosols on clouds that interact directly with radiation. Aerosols can change the size of the water droplets and ice crystals in the lower atmosphere and can increase their number. This can lead to increased reflection of solar radiation by the clouds and a tendency to cool the surface. These processes are complex. Their intensities vary in space and depend on the local vertical distribution of aerosols, their size, and their nature. Many uncertainties remain about the effects of aerosols on the climate.

4.6 Effects of clouds on radiation

When present, clouds are the most influential atmospheric component in attenuating solar radiation. A cloud is defined as a visible mass suspended in the atmosphere, consisting of a large quantity of water droplets, or ice crystals, or both.

Table 4.1 The ten genera of clouds with their corresponding altitude range and phase

Cloud genera	Altitude range	Phase
Stratus, nimbostratus, stratocumulus	0–2 km	Liquid, water droplets
Altocumulus, altostratus	2–6 km	Liquid, water droplets
Cirrus, cirrostratus, cirrocumulus	5–13 km	Solid, ice crystals
Cumulus	Fairly large vertical extension	Liquid, water droplets
Cumulonimbus	Very large vertical extension. The base is at a few hundreds of m up to 5 km above the ground. The top reaches the tropopause, even higher	Liquid, water droplets and solid, ice crystals in the upper layer

Clouds exhibit a great variety for both their optical and geometric properties. Meteorologists have developed very elaborate classifications of different clouds, mainly based on visual observation of clouds and less on cloud physics, although the two aspects are linked. The International Cloud Atlas[7] published in 2017 by the World Meteorological Organization details the classification system that is used worldwide to ensure consistency of observer readings. I limit myself here to a brief description of the clouds, which I believe is useful for a better understanding of the solar radiation arriving on the ground. Clouds appear mostly in the troposphere. The classification distinguishes between levels, a level being a layer of the atmosphere in which a family of clouds appears most frequently. The atlas describes ten main groups, called genera, of which the three main are:

- stratus that are generally low gray clouds that tend to develop horizontally, like a cover;
- cirrus that are detached clouds at high altitude, in the form of white, delicate filaments or white or mostly white patches or narrow bands usually thin;
- cumulus that are generally low detached clouds with vertical development, i.e., a fairly large vertical extension.

Table 4.1 lists the ten genera with the corresponding altitude range, as well as the phase, either liquid and composed of water droplets or solid and composed of ice crystals, since the phase has a significant influence on the effect of the cloud on solar radiation.

Regarding the radiative transfer, the optical properties of clouds are considered rather than their visual aspects. The absorption of radiation by water droplets and ice crystals is very low, except in the near-infrared range due to water vapor (see Figure 4.2). Water droplets and ice crystals are bodies that scatter solar photons. Scattering, and therefore the attenuation of radiation, by a cloud is very important

---

7 International Cloud Atlas Manual on the Observation of Clouds and Other Meteors (WMO-No. 407), Edition 2017, available in digital format for free at https://cloudatlas.wmo.int/home.html.

because of the density of the scattering bodies within the cloud. The optical depth of the clouds can vary from 0 to 100 or more. For a sun at vertical ($\theta_S=0°$), an observer will no longer see the sun clearly when the optical depth of the cloud is around 3. In this case, the transmittance is around 0.05; i.e., 95 % of the flux is not transmitted by the cloudy column.

The optical depth of the cloud is linked to its vertical extension, and therefore to its geometric thickness, since the attenuation is linked to the length of the path of the rays in the cloud. It is also dependent on the phase of the cloud and the size of the water droplets or ice crystals. Generally, cirrus clouds do not attenuate much the radiation and have small optical depths. In contrast, cumulonimbus greatly attenuates radiation; darkness can prevail under this cloud, even at noon.

The size of the water droplets is around 10 μm, and that of ice crystals is 20 μm. The cross section of these scatterers considered as spheres is approximately 60 and 120 μm, respectively, which is large compared to the wavelengths of solar radiation, since the latter are between 0.2 and 4 μm for 99 % of the extraterrestrial total radiation. In these conditions, the scattering is non-selective; i.e., it affects all wavelengths uniformly, except for the longest in the infrared. This explains why the clouds appear white, or gray, to the observer on the ground because all the wavelengths of human vision are scattered in an identical manner.

The role of clouds is not limited to low absorption and high scattering. Indeed, they have a generally high albedo, i.e., a large capability of reflecting the radiation, and return the radiation toward space with a notable effect on the radiative balance and the climate. They also backscatter downward the solar radiation that was reflected by the ground (see Figure 4.12 below), thus increasing the diffuse component of the radiation received by a horizontal plane on the ground.

If the cloud cover is sparse, the sun can be seen by an observer on the ground between two clouds if no cloud intercepts the radiation in the direction of the sun. A horizontal surface on the ground can then receive an irradiance whose direct component is that of a clear sky and whose diffuse component can be greater than that of clear sky because of the multiple scattering by the neighboring clouds (see Figure 4.12 below). In certain circumstances, the vertical faces of clouds can scatter photons in such a way that they are concentrated on the horizontal plane on the ground. It is then possible to measure on this plane irradiance greater than the extraterrestrial one for periods of up to 1 h. This can also happen when the sun is low on the horizon in the presence of a layer of thick clouds that do not obscure the sun. The surface then receives the rays coming from the sun plus those that are reflected by the underside of the clouds.

4.7 The paths of the sun rays in the atmosphere

Scattering can modify the trajectory of incident solar rays, unlike absorption. Strictly speaking, the description of photon paths in the atmosphere must be done using probabilities. I voluntarily simplify the speech here and I adopt a geometric description, certainly less rigorous but easier to explain and make understood. This simplification has no impact on the rest.

Figures 4.11 and 4.12 are schematic representations of the different paths of the solar rays as they pass through the atmosphere from top to bottom. As with the following

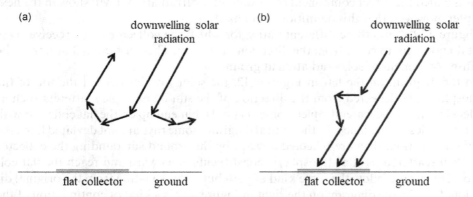

Figure 4.11 Schematic view of different paths of the downwelling solar rays. (a) Solar rays do not reach the ground. (b) Solar rays reaching the flat collector are parallel and seem to come from the direction of the sun.

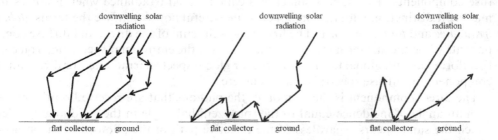

Figure 4.12 Schematic view of the different paths of the downwelling solar rays received by a flat collector at ground, showing various contributions to the diffuse component of the radiation at ground.

diagrams, these are valid for a particular wavelength as well as for integrals on all or part of the solar spectrum.

In the diagram on the left in Figure 4.11, the solar rays incident at the top of the atmosphere are deviated from their original direction by multiple scattering due to the constituents of the atmosphere (molecules, aerosols, water droplets, ice crystals). Their routes are such that they do not reach the collector plane on the ground; they therefore do not contribute to the irradiance received by the plane and are a net loss from this point of view.

In the diagram on the right in Figure 4.11, the solar rays incident at the top of the atmosphere are not deviated from their original direction, that of the sun, by the multiple scattering. I remind you that in the atmosphere, the probability of having a forward scattering, i.e., in the same direction as the incident, is generally greater than for the other directions. There can also be multiple scattering that deviates rays from the original direction, before bringing them back into the original beam. The incident rays on the collector plane are parallel, and all seem to come from the direction of the sun, the latter appearing at a solid angle with an apparent diameter of about $32'\pm0.5'$ of arc.

This is called the direct component of irradiance, or irradiation. I will show in the next chapter, however, that this definition is not as clear.

Figure 4.12 shows three different paths, for which the collector plane receives rays that do not seem to come from the direction of the sun. These are contributions to the diffuse component of solar radiation at ground.

In the diagram on the left in Figure 4.12, the solar rays incident at the top of the atmosphere are deflected from the direction of the sun by multiple scatterers such as molecules, aerosols, water droplets, or ice crystals, but end up on the flat collector with various angles of incidence. In the central diagram, some rays are not deviated from the original direction; they are reflected upward by the ground surrounding the collector and then scattered by the atmospheric constituents downward and reach the flat collector. Other rays follow the same kind of path but after deviating from the original direction. Finally, the diagram on the right in Figure 4.12 is a kind of continuation of the diagram on the right in Figure 4.11 showing the direct component. The "direct" rays are reflected by the flat collector itself; some of them are backscattered by the various atmospheric constituents and return to the collector with various angles of incidence.

The irradiance received by the horizontal collector is the sum of its direct and diffuse components. This sum is sometimes called global irradiance when it comes to making a distinction with its components. Be careful not to confuse the terms *global irradiance* and *total irradiance*. The first term is the sum of the direct and diffuse components while the last term means the integral over the entire solar spectrum. You can therefore deal with global total irradiance, or global spectral irradiance, or direct total component, or diffuse spectral component, etc.

The direct component is the sum of all the photons that are received on the plane with an angle of incidence equal to the solar zenithal angle in the solid angle under which the sun appears regardless of the effective path of the photons in the atmosphere. The diffuse component is the sum of all the other photons received by the horizontal collector.

4.8 Summary of contributions of atmospheric constituents to attenuation of radiation

All the constituents of the atmosphere contribute in varying degrees to the attenuation of solar radiation. In the stratosphere, the main phenomena are the absorption of X-rays and ultraviolet rays and the scattering of short wavelengths in purple and blue. Radiation with wavelengths less than 300 nm is absorbed by ozone. The wavelengths of the radiation reaching the ground are mainly between 300 and 4000 nm. As the solar rays enter the atmosphere, the longer wavelengths are more and more attenuated. Variations in the concentrations of gases and aerosols, as well as in their optical properties, cause the attenuation of the atmosphere to vary over time and in geographic space with sometimes very short time and space scales. It is therefore difficult to estimate very accurately the effects of the atmosphere on solar radiation.

Table 4.2 schematically summarizes the contributions of the various constituents of the atmosphere to the attenuation of the solar radiation incident at the top of the atmosphere, during its downward path toward the ground. The latter plays an important role; it contributes to the diffuse component with generally a strong spectral dependence. This is the subject of the following chapter.

Table 4.2 Summary of the contributions of the various constituents of the atmosphere to the attenuation of solar radiation during its downward path to the ground

Constituent	Scattering	Absorption
Ozone	Can be neglected	Strong for λ less than 300 nm, otherwise negligible
Gases (other than ozone and water vapor)	Strong (in λ^{-4}). Increases as the wavelength decreases. Negligible for λ greater than 700 nm	Weak
Water vapor	Can be neglected	Significant to very strong when λ is greater than 650 nm
Aerosols	Can be strong, dependence on $\lambda^{-\alpha}$ with $-1 \leq \alpha \leq 4$	Weak
Clouds	Strong, weak dependence on λ	Weak

λ is the wavelength.

Chapter 5

Ground reflection

At a glance

The solar radiation reaching the ground is reflected by the ground. The reflected radiation can contribute after multiple scattering to the irradiance received by a horizontal plane.

The reflection of the ground strongly depends on the material, its shape, its surface properties, including its roughness, the wavelength of the incident radiation, and the angle of incidence. It is rarely known with great accuracy. It can be characterized by the reflection factor or the bidirectional reflectance distribution function. These two quantities are related and describe how a beam of parallel rays is reflected according to the directions of illumination and observation.

The reflection factor or reflectance quantifies the effectiveness of a body in reflecting solar radiation. The greater the reflection factor, the greater the reflected radiation. The bidirectional reflectance distribution function is an intrinsic property of the reflecting body. It is often approximated by an isotropic part, which is the main contributor under most conditions, and an anisotropic part.

Albedo is a quantity that synthesizes the combined effects of the attenuation of the solar radiation by the atmosphere and the reflection of the ground. Black-sky albedo and white-sky albedo are other quantities to describe the reflection of natural surfaces. The black-sky albedo is that which would be obtained if the downwelling irradiance were composed only of its direct component. The white-sky albedo is that which would be obtained with isotropic illumination of the surface.

The reflection of the ground presents significant spectral variations. The spectral distributions are various depending on the surfaces and weather conditions. The reflection properties and the associated quantities can be defined for a particular wavelength or over spectral intervals or over the entire spectrum.

Part of the radiation reaching the ground is reflected, part of which is backscattered toward the ground. The central and right diagrams of Figure 4.12 of Chapter 4 highlighted the importance of the reflection by the ground and the collector in the diffuse component of the irradiance. This chapter addresses the reflection of the solar radiation by the ground. Several quantities, namely the reflection factor, the reflectance,

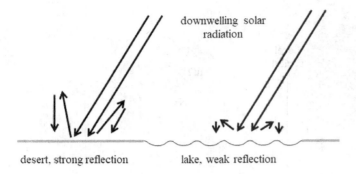

desert, strong reflection   lake, weak reflection

Figure 5.1 Schematic view of the reflection on the Thartar Lake and surroundings and the contribution of the ground to the diffuse component.

the bidirectional reflectance distribution function, and the albedo, are defined that describe how the sun rays are reflected by the ground. Several examples of the spectral distribution of the reflectance are given.

First of all, to better enlighten you, allow me, reader, to illustrate the importance of the phenomenon with the example of Thartar Lake, in Iraq, which is not at all an exceptional case. Figure 5.1 is a schematic view of this lake that is surrounded by a large desert area of erg type. Let me assume a spatially homogeneous atmosphere, without cloud, as well as a weak reflection of water (except for specular reflection). The direct component of the irradiance received by a plane at the surface of the water is large: It typically represents 70 % of the global irradiance. About 10 km inland, reflection of the solar radiation by rocky and sandy soil is more important than on the lake and contributes more to the diffuse component. As the direct component is the same, the irradiance received on a flat collector located in the erg is greater than on the lake. The reality is more complex, notably due to the presence of clouds, but this rapid calculation may explain a relative difference in irradiance observed in the order of 5 %.

5.1 Reflection factor – its spectral variations

Generally, when the rays arrive on the collecting plane, a part is absorbed, a part is possibly transmitted if the plane is not opaque, and the rest is reflected. The reflection strongly depends on the material, its shape, its surface properties, including its roughness, the wavelength of the incident radiation, and the angle of incidence. The reflection properties and the associated quantities can be defined for a particular wavelength or over spectral intervals or over the entire spectrum.

5.1.1 Reflection factor

Consider an incident flux in the direction of incidence or illumination (θ, ψ) received by a collector plane, where θ is the angle of incidence and ψ is the corresponding azimuth. Figure 5.2 depicts the reflection in the direction of observation (θ', ψ'), where θ' is the angle of reflection and ψ' is the corresponding azimuth.

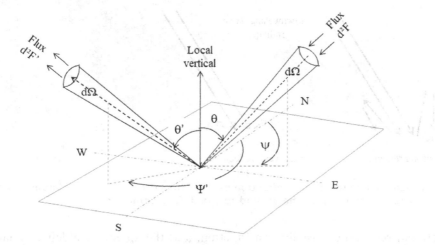

Figure 5.2 Diagram illustrating the reflection. The incident flux arrives from the direction of illumination (θ, ψ). It can be reflected in several directions, including the direction (θ', ψ'), called here observation direction.

The reflection factor $r(\theta, \psi, \theta', \psi')$ is defined as the ratio of the radiance $L'(\theta', \psi')$ reflected in the direction of observation to the incident radiance $L(\theta, \psi)$:

$$r(\theta, \psi, \theta', \psi') = L'(\theta', \psi')/L(\theta, \psi) \tag{5.1}$$

The reflection factor is also called directional reflectance and is dimensionless. A perfectly reflecting material has a reflection factor equal to 1 in all directions.

Specular reflection is a special case of reflection. It occurs when the energy reflected is only in the opposite direction to the direction of illumination:

$$r(\theta, \psi, \theta', \psi') = 0 \text{ everywhere, except where} \theta' = \theta \text{ and} \psi' = -\psi \tag{5.2}$$

Specular reflection occurs on a very smooth surface that reflects like a mirror. A common case is that of calm, fairly clear waters more than a few meters deep, such as oceans and lakes. Solar radiation is reflected in the direction opposite to the illumination direction. When the sea state gets rougher due to the wind for example, the reflective small facets are more numerous and more randomly orientated, increasing the roughness of the water surface. Under these conditions, the reflection of water departs from the specular case, and diffuse reflection, that is, reflection in several directions, if not in all directions, becomes more important.

The case of water is symptomatic of the influence on the reflection of the roughness of the surface of the reflecting material with respect to the incident wavelength. The rougher the surface, i.e., the more it offers asperities whose size is greater than the wavelength of the incident radiation, the more diffuse the reflection. If the sizes are smaller, the reflection is closer to the specular reflection. The vast majority of natural surfaces have a predominantly diffuse reflection. A perfectly diffusing material reflects the flux received in all directions, but not necessarily equally. Lambertian reflection is a special case of diffuse reflection, for which the reflected radiance $L'(\theta', \psi')$ is orthotropic, that is to say, that it is the same angularly in all directions of the hemisphere above the horizontal collector plane.

The reflection of natural surfaces is not easy to know with accuracy because the reflection depends on their exact composition, their optical properties, and their roughness, which can vary over time. For example, snow is less reflective if it is very wet or soiled. It offers mainly diffuse reflection, but there are cases where it can be specular. The same applies to the vast salt lakes often dried up like the sebkhas, also called playas, which are very shallow floodable depressions where the soils are salty and the vegetation thin and sparse, or the salted remains of great lakes like Lake Bonneville in the United States, or the Aral Sea in Asia. When a thunderstorm occurs, the sebkha is flooded. It reflects less and the reflection is mainly diffuse. Then gradually, the water evaporates leaving the salt crust that reflects much more and anisotropically. Another example is given by the deciduous forests that reflect less in winter when the trees are leafless than in summer when they bear new foliage. But this can be different if in winter the ground is covered with thick, well-reflecting snow. Let me give a last example taken from cultivated soils. When sowing after plowing, a field looks like bare soil with a high reflection factor and offers diffuse reflection. As plants grow, the reflection factor changes in value and angular distribution.

When the reflective body has a certain volume and is not opaque, like a tree, for example, the reflection takes place on several levels, for example, at the level of the leaves, the trunk, and the ground. This is called volume reflection.

5.1.2 Examples of spectra of reflection factor

I mentioned above that the reflection depends on the roughness of the reflecting surface with respect to the wavelength of the incident radiation. Consequently, the reflection factor or directional reflectance r presents spectral variations. It can be defined for a particular wavelength λ or over spectral intervals $[\lambda_1, \lambda_2]$ or over the entire spectrum. Be careful, the reflection factor is a ratio of radiances that depend on λ. Consequently, the reflection factor $r(\lambda_1, \lambda_2)$ is not the integral of $r(\lambda)$ between λ_1 and λ_2. The definitions of spectral reflection factor are as follows:

$$r(\lambda, \theta, \psi, \theta', \psi') = L'(\lambda, \theta', \psi') / L(\lambda, \theta, \psi) \tag{5.3}$$

$$r(\lambda_1, \lambda_2, \theta, \psi, \theta', \psi') = L'(\lambda_1, \lambda_2, \theta', \psi') / L(\lambda_1, \lambda_2, \theta, \psi) \tag{5.4}$$

The spectral reflection factor is dimensionless.

The two following graphs exhibit spectra of reflectance between 350 and 2500 nm for certain natural bodies, chosen at random and only for illustration purposes. The graphs were made from spectra available in digital form from the USGS agency in the USA.[1]

---

1 Spectral Library Version 7 – Base Spectra (splib07a), available at crustal.usgs.gov/speclab/Query-All07a.php. [Original measurements made using lab, field and imaging spectrometers. Kokaly R. F., Clark R. N., Swayze G. A., Livo K. E., Hoefen T. M., Pearson N. C., Wise R. A., Benzel W. M., Lowers H. A., Driscoll R. L., Klein A. J., 2017. USGS Spectral Library Version 7: U.S. Geological Survey Data Series 1035, 61 p. doi:10.3133/ds1035.

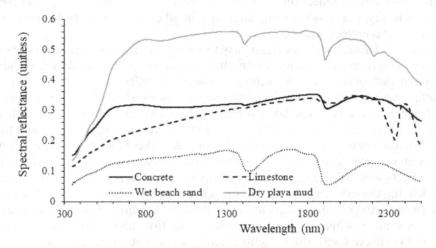

Figure 5.3 Examples of reflectance spectra between 350 and 2500 nm for concrete, limestone, wet beach sand and dry playa mud. Graphs made from spectra kindly provided by the federal agency USGS of the United States.

Figure 5.3 shows examples of reflectance spectra for concrete, limestone, wet beach sand, and dried mud in a sebkha or playa. The spectral curve for concrete is fairly flat, around 0.3, with lower values, down to 0.15 at wavelengths less than 600 nm. The reflectance also decreases beyond 2100 nm. The curve for limestone increases fairly regularly with the wavelength, from 0.10 to about 0.30 for 2200 nm, then presents some indentations. The reflectance for wet beach sand is fairly small and less than 0.15. The spectral curve increases fairly regularly with the wavelength, up to around 1300 nm, then exhibits a succession of hollows and bumps. The dried mud of sebkha exhibits a minimum in reflectance of 0.15 at 350 nm and increases rapidly with the wavelength, reaching a high plateau of 0.55 around 700 nm. The reflectance decreases when the wavelength is greater than 1800 nm.

Figure 5.4 exhibits the spectral distributions for some plants: grass, poplar, and conifer. First of all, note that whatever the plant, the reflectance is weak and less than 0.04 at 350 nm. The reflectance then grows, more or less quickly and regularly.

The spectral curves for the grass are fairly close to each other, although the reflectances are different. They have roughly the same hollows and bumps. The reflectance slowly decreases irregularly from 1300 nm. The curve for poplar shows a sharp increase in reflectance around 700 nm to 0.48, then a plateau and an irregular decrease from 1300 nm with the same hollows and bumps as the grasses. The reflectance for the conifer is very low and less than 0.02 up to 700 nm and then shows a strong increase up to 0.2 and less up to 0.25 at 1100 nm, then a very irregular decrease that only follows approximately the hollows and bumps of the other plants discussed here.

5.2 Bidirectional reflectance distribution function (BRDF)

The reflection depends on the directions of illumination (θ, ψ) and observation (θ', ψ'). As previously, let $L(\theta, \psi)$ be the radiance incident on a horizontal plane in the direction

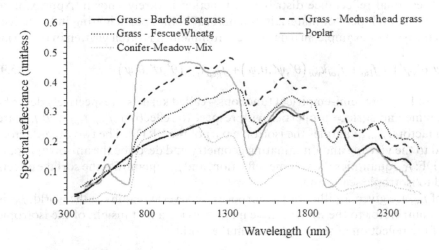

Figure 5.4 Examples of reflectance spectra between 350 and 2500 nm for a few types of grass and trees. Graphs made from spectra kindly provided by the federal agency USGS.

of illumination (θ, ψ). The radiance $L'(\theta', \psi')$ reflected in the solid angle Ω' centered on the direction of observation (θ', ψ') is:

$$L'(\theta',\psi')= \int_{\Omega'} \rho(\theta,\psi,\theta',\psi')L(\theta,\psi)\cos(\theta)d\omega \qquad (5.5)$$

where ρ is such as:

$$\rho(\theta,\psi,\theta',\psi')= p_{reflected}(\theta,\psi,\theta',\psi')/\cos(\theta') \qquad (5.6)$$

and $p_{reflected}(\theta, \psi, \theta', \psi')$ is the probability for an incident photon incident to be reflected in the direction (θ', ψ').

The function ρ is called the bidirectional reflectance distribution function, abbreviated as BRDF. Its unit is sr^{-1}. It is positive and verifies the principle of reciprocity:

$$\rho(\theta,\psi,\theta',\psi')= \rho(\theta',\psi',\theta,\psi) \qquad (5.7)$$

The function ρ is such that it is conserving energy; i.e., the energy reflected in all directions by a reflecting plane illuminated from the direction (θ, ψ) cannot be greater than the energy incident in this direction. The two energies can be equal if the reflecting plane does not absorb energy. The relationship between the reflection factor r and ρ is:

$$r(\theta,\psi,\theta',\psi')= \rho(\theta,\psi,\theta',\psi')\cos(\theta) \qquad (5.8)$$

Compared to the reflection factor r, the BRDF ρ is normalized with respect to the angle of incidence θ. Like r, the BRDF ρ depends on the wavelength λ. Be careful, the BRDF $\rho(\lambda_1, \lambda_2)$ integrated over a spectral range $[\lambda_1, \lambda_2]$ is not the integral of $\rho(\lambda)$ between λ_1 and λ_2.

The bidirectional reflectance distribution function is rarely known. Approximate laws have been developed for materials and natural bodies participating in the reflection of solar rays. For example, in a natural environment, ρ is often written in the form:

$$\rho(\theta,\psi,\theta',\psi') = f_{iso} + f_{vol}k_{vol}(\theta',\psi',\theta,\psi) + f_{geo}k_{geo}(\theta',\psi',\theta,\psi) \tag{5.9}$$

where k_{vol} and k_{geo} are semi-empirical functions – called kernels – respectively describing the volume and surface properties with respect to reflection. f_{iso}, $f_{vol,}$ and f_{geo} are weighting factors. f_{iso} quantifies the isotropic part of the BRDF. The two other factors are related to the viewing and illuminating geometry and describe the anisotropic part of the BRDF: f_{vol} quantifies the volume reflection, and f_{geo} quantifies the surface effects compared to isotropic scattering.

Maps of f_{iso} are given for illustrative purposes for several regions in the world. f_{iso} is the major contributor to the BRDF; these maps provide a first insight of the isotropic part r_{iso} of the reflection of solar radiation in the world:

$$r_{iso}(\theta) = f_{iso}\cos(\theta) \tag{5.10}$$

The following maps are drawn from the data set of f_{iso}, $f_{vol,}$ and f_{geo} parameters assembled by my former Mines ParisTech institute in France.[2] Maps are drawn up for 4 months: January, April, July, and October, in order to offer an overview of the variations of f_{iso} (in sr^{-1}) during the year.

Figure 5.5 exhibits the Southeast Asia and Australia region. The ocean is in black because f_{iso} is very small, about 0.02. The islands at the top of the map are mainly covered with broadleaf evergreen forests and agricultural areas. They exhibit weak f_{iso}, which do not change much during the year. More spatial contrast can be seen over Australia where the central part made of arid and sandy desert is brighter than its surroundings. The eastern part offers the smallest values of f_{iso} and is the place of agricultural activities; it shows small variations in f_{iso} during the year. The brightest area corresponds to the Lake Eyre basin that is a very large arid and semi-arid area; f_{iso} is great all year round, about 0.7. In New Zealand, large f_{iso} due to snow can be spotted in the Southern Alps in winter (July, October).

Other regions of the world are drawn in Figures 5.6–5.10. I do not comment them. If you are familiar with a region, you will recognize the most salient geographic features, such as mountains or river basins. Ocean and other water bodies exhibit very small f_{iso} and are in black. Snow may cover plains and mountains in winter with high f_{iso} and may disappear in spring. Equatorial and tropical forests offer small variations in f_{iso} all year round.

5.3 Albedo

Albedo is a quantity that concerns reflection, but also the backscattering of a diffusing and absorbing medium, or both. An example is given by the assembly made up of the

2 Blanc P., Gschwind B., Lefèvre M., Wald L., 2014. Twelve monthly maps of ground albedo parameters derived from MODIS data sets. In Proceedings IGARSS 2014, 14–18 July 2014, Quebec, Canada, US-BKey, 3270–3272. The data set is freely available at www.oie.mines-paristech.fr/Valorisation/Outils/AlbedoSol/

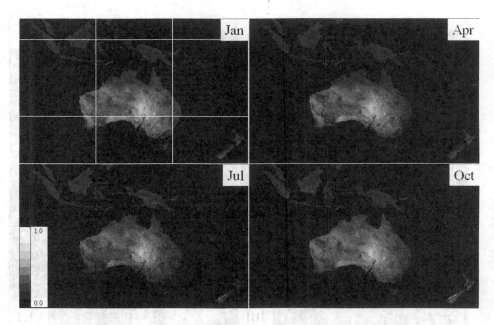

Figure 5.5 Maps of the parameter f_{iso} describing the isotropic part of the bidirectional reflectance distribution function in Southeast Asia and Australia. Upper left: A grid every 30° in latitude and longitude is superimposed. Drawn from the data set kindly provided by Mines ParisTech of France.

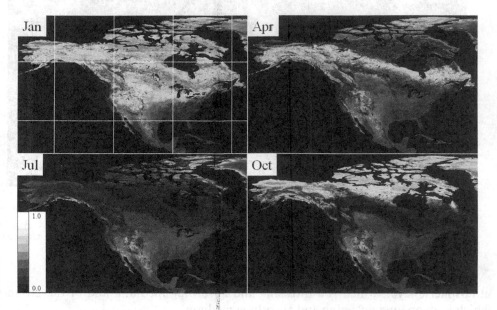

Figure 5.6 Maps of the parameter f_{iso} describing the isotropic part of the bidirectional reflectance distribution function in North America. Upper left: A grid every 30° in latitude and longitude is superimposed. Drawn from the data set kindly provided by Mines ParisTech of France.

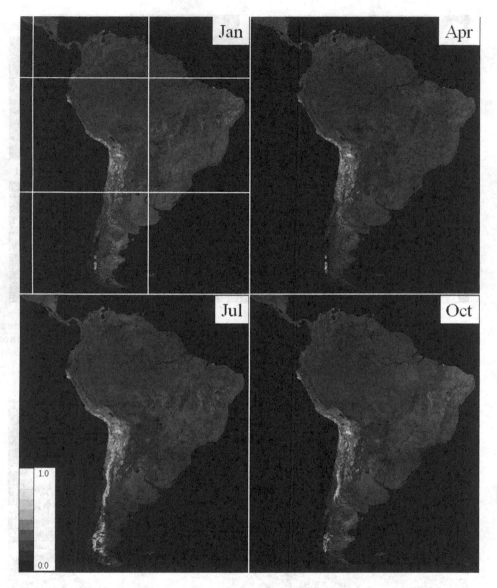

Figure 5.7 Maps of the parameter f_{iso} describing the isotropic part of the bidirectional reflectance distribution function in South America. Upper left: A grid every 30° in latitude and longitude is superimposed. Drawn from the data set kindly provided by Mines ParisTech of France.

atmospheric layer, which is a diffusing and absorbing medium, and of the ground, which is an opaque reflecting and absorbing medium.

Assume a reflective body or a scattering–absorbing layer, or all of them. Let $E_{incident}$ be the irradiance illuminating this body or this layer, measured just above, and $E_{reflected}$ be the upwelling irradiance measured just above. $E_{reflected}$ is the sum of the

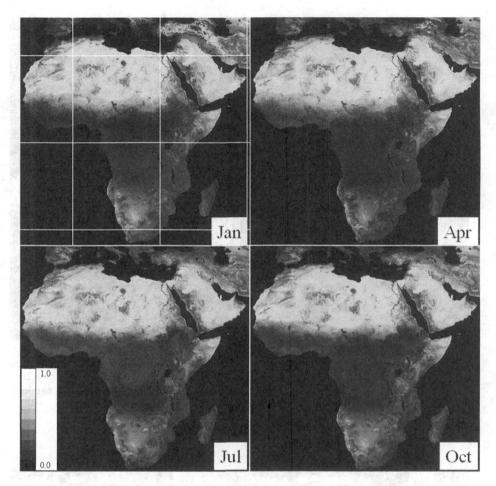

Figure 5.8 Maps of the parameter f_{iso} describing the isotropic part of the bidirectional reflectance distribution function in Africa. Upper left: A grid every 30° in latitude and longitude is superimposed. Drawn from the data set kindly provided by Mines ParisTech of France.

reflected and backscattered irradiances. The albedo A is defined as the ratio of $E_{reflected}$ to $E_{incident}$:

$$A(\theta,\psi) = E_{reflected} / E_{incident} \qquad (5.11)$$

For example, if $E_{downwelling}$ indicates the downwelling irradiance received on the ground, and if $E_{upwelling}$ indicates the reflected irradiance, the albedo of the ground A_{ground} is:

$$A_{ground}(\theta,\psi) = E_{upwelling} / E_{downwelling} \qquad (5.12)$$

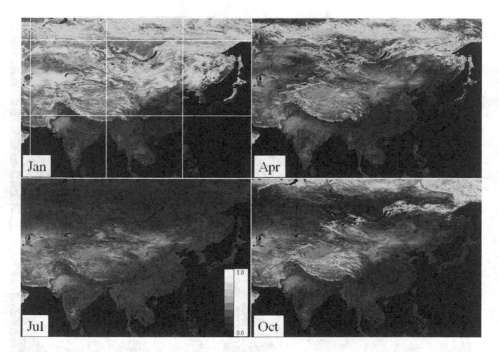

Figure 5.9 Maps of the parameter f_{iso} describing the isotropic part of the bidirectional reflectance distribution function in Asia. Upper left: A grid every 30° in latitude and longitude is superimposed. Drawn from the data set kindly provided by Mines ParisTech of France.

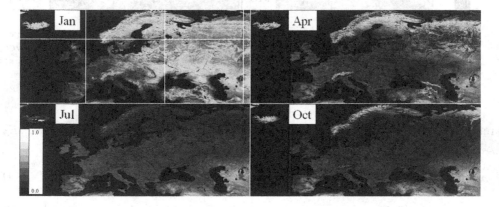

Figure 5.10 Maps of the parameter f_{iso} describing the isotropic part of the bidirectional reflectance distribution function in Europe. Upper left: A grid every 30° in latitude and longitude is superimposed. Drawn from the data set kindly provided by Mines ParisTech of France.

Albedo is a dimensionless quantity. A perfectly reflecting body has an albedo equal to 1; a perfectly absorbing body – a black body – has an albedo equal to 0. In the case of a non-reflecting scattering medium, an albedo of 1 corresponds to a medium that scatters backward all the photons it receives; an albedo of 0 may correspond to a fully absorbing medium, or one without backscattering, or a combination of the two.

The albedo presents spectral variations. It can be defined for total irradiance: total albedo, or for a range of wavelengths $[\lambda_1, \lambda_2]$ or a particular wavelength λ. Be careful the albedo $A(\lambda_1, \lambda_2)$ is not the integral of $A(\lambda)$ between λ_1 and λ_2. The definitions of spectral albedo are as follows:

$$A(\lambda) = E_{upwelling}(\lambda) / E_{downwelling}(\lambda) \tag{5.13}$$

$$A(\lambda_1, \lambda_2) = E_{upwelling}(\lambda_1, \lambda_2) / E_{downwelling}(\lambda_1, \lambda_2) \tag{5.14}$$

What is the difference between the albedo and the reflection factor in the case of an opaque reflecting medium? The reflection factor is a ratio of radiances, while the albedo is a ratio of irradiances, which are hemispherical integrals of radiances.

Natural surfaces present various albedos with varied spectral distributions. Take some typical cases of total albedo. Except in the case of specular reflection, a body of water deep enough so that the bottom does not interfere has an albedo very close to 0, typically 0.02, or even a little more depending on the agitation of the surface of the water. A forest has an albedo of about 0.15 for conifers and 0.20 for deciduous trees. A cultivated field has a high albedo when it has just been sown, around 0.25, and this albedo can drop to 0.15 during growth when the plants hide the soil. Depending on its size and humidity, the sand has an albedo between 0.20 and 0.45. Snow offers a variety of albedo, from 0.4 to 0.9, depending on whether it is cool, wet, or thick. If it is soiled, its albedo is weaker.

The notions of black-sky albedo and white-sky albedo may be used to describe the reflection of natural surfaces. The black-sky albedo $A_{black-sky}$ is the albedo that would be obtained if the downwelling irradiance $E_{downwelling}$ were composed only of its direct component E_{direct}. It is also called directional hemispherical reflectance or direct albedo. The white-sky albedo $A_{white-sky}$ is that which would be obtained with isotropic illumination of the surface; it is also called bihemispherical reflectance, or hemispherical albedo.

If $E_{diffuse}$ indicates the diffuse component of $E_{downwelling}$, then the albedo A is given by one of the following equivalent relationships:

$$A = \left(E_{direct} / E_{downwelling} \right) A_{black-sky} + \left(E_{diffuse} / E_{downwelling} \right) A_{white-sky} \tag{5.15}$$

$$\text{or} \quad A = \left(E_{direct} / E_{downwelling} \right) A_{black-sky} + \left(1 - E_{direct} / E_{downwelling} \right) A_{white-sky} \tag{5.16}$$

$$\text{or} \quad A = \left(1 - E_{diffuse} / E_{downwelling} \right) A_{black-sky} + \left(E_{diffuse} / E_{downwelling} \right) A_{white-sky} \tag{5.17}$$

Given these relationships, it is clear that albedo is not an intrinsic property of the surface. It depends on the downwelling irradiance $E_{downwelling}$ and its components $E_{diffuse}$ and E_{direct}, in particular on the diffuse fraction $F_{diffuse}$, with $F_{diffuse} = E_{diffuse} / E_{downwelling}$, or in an equivalent manner, on the direct fraction F_{direct}, with $F_{direct} = E_{direct}/E_{downwelling}$. The previous relationships can be rewritten as:

$$A = \left(1 - F_{diffuse}\right) A_{black\text{-}sky} + \left(F_{diffuse}\right) A_{white\text{-}sky} \tag{5.18}$$

$$A = \left(F_{direct}\right) A_{black\text{-}sky} + \left(1 - F_{direct}\right) A_{white\text{-}sky} \tag{5.19}$$

Assume a surface having black-sky $A_{black\text{-}sky}$ and white-sky $A_{white\text{-}sky}$ albedos. The albedo A takes different values depending on the diffuse (or direct) fraction. For example, if the sky is cloudless, the diffuse fraction is about 0.3 and the ground albedo is:

$$A = 0.7 A_{black\text{-}sky} + 0.3 A_{white\text{-}sky} \tag{5.20}$$

But if the sky is overcast, in which case the diffuse fraction is equal to 1, the ground albedo is now:

$$A = A_{white\text{-}sky} \tag{5.21}$$

Most of the discussions in the above paragraphs have focused on reflective bodies. The albedo of natural scattering media should not be overlooked. The atmosphere scatters the reflected radiation backward to the ground as shown in Figure 4.12, even under clear skies. An albedo of the atmosphere can be defined which characterizes this backscattering by all the atmospheric layers and their constituents. The more turbid the atmosphere, the greater the atmospheric albedo under clear-sky conditions.

Clouds are other natural scattering media. They have a wide variety of albedo, ranging from 0.2 or less to 0.9. As it comes to scattering, there is a relationship between the optical depth of a cloud and its albedo. The greater the optical depth of the cloud, the greater the albedo.

Climatologists have a special interest in the planetary albedo of the Earth, i.e., of atmosphere and ground. The planetary albedo is the ratio of the upwelling total irradiance at the top of the atmosphere resulting from reflection and backscattering by the atmosphere and the ground to the extraterrestrial total irradiance E_0, taken for the whole of the Earth, and not just a particular geographic point.

The planetary albedo is worth around 0.3. General cooling of the atmosphere can lead to an increase in the area covered by ice, which has a high albedo. As a result, solar radiation is more reflected and less absorbed by the ground, increasing the planetary albedo. On the contrary, a global warming can lead to a decrease in the surfaces covered by the ice, less reflection, greater absorption by the soil, and a decrease in planetary albedo. This decrease can itself lead to an increase in the temperature of the atmosphere and thus reinforce the phenomenon of warming.

Chapter 6

Solar radiation received at ground level

At a glance

The solar radiation, total or spectral, received on a horizontal plane on the ground, is the sum of its direct and diffuse components. The direct component can be defined as the sum of all the photons that are received on the plane in the solid angle under which the sun appears. Its exact definition depends on the field of application, even on the measuring instruments, and must be specified. The diffuse component is the sum of all the other photons received by the plane. If the plane is tilted, it also receives part of the radiation reflected from the neighboring ground, which is called the reflected component. When it is necessary to distinguish between the components and their sum, the latter is called global radiation.

The clearness index is a normalization of the radiation received on the ground by the extraterrestrial radiation. It quantifies the attenuation due to the atmosphere. The clearer the sky, the greater the clearness index. At midlatitudes, the clearness index is around 0.45 on an annual average. The clear-sky index is a normalization of the radiation actually received on the ground by the radiation that would be received on the ground in the absence of cloud. It quantifies the attenuation due to clouds only. It is also known as the cloud modification factor.

Except in special cases of concentration of solar rays on the sides of the clouds, the radiation received on the ground under clear-sky conditions is greater than that received under cloudy conditions. Attenuation of solar radiation, total or spectral, by clouds plays a dominant role, except in the absorption lines specific to gases. The clouds induce a more or less marked darkening of the sky for the observer on the ground. Clouds and their properties are very variable in time and space, which induces more or less significant variations in radiation on the ground in time and space.

It is fairly easy to calculate the incident solar radiation at the top of the atmosphere. The calculation is much more difficult for the radiation received on the ground because many variables must be known at any time and any place, all varying in time and space. The most influential variables on radiation are date, time, geographic location, the properties of the constituents of the clear and cloudy atmosphere, and the reflective properties of the ground. The calculation of the radiation received on the ground under all sky conditions can be simplified by first calculating the radiation in clear-sky conditions, and then by calculating the attenuation of this clear-sky radiation by the clouds.

I established in Chapter 3 the incident solar radiation at the top of the atmosphere, or extraterrestrial radiation, which, I recall, constitutes in most cases an upper limit of the irradiance or irradiation received on the ground. I described in Chapter 4 the influence of the atmosphere on the solar radiation during its crossing toward the ground, with the absorption by gases, multiple scattering by the atmospheric constituents, and reflections by the ground discussed in Chapter 5. In this chapter, I combine this diverse knowledge to present the solar radiation, total and spectral, received on the ground by a horizontal or tilted plane, such as a natural slope or a rooftop. This chapter deals with the total radiation, while Chapter 7 is dedicated to the spectral distribution.

The solar radiation, total or spectral, received on a horizontal plane on the ground, is the sum of its direct and diffuse components. There is also a reflected component due to the reflection of radiation from the surrounding ground, if the receiving surface is tilted or in case of relief. The sum of these components is called global radiation. I begin this chapter with the definition of these components and their description.

I then define other quantities, such as the clearness index and the diffuse fraction, which are very practical standardized quantities for comparing irradiances and irradiations between different places of the world and different seasons.

I insist in this chapter on the predominant role of clouds in the attenuation of solar radiation, total or spectral. Clouds, their presence, and their optical and geometric properties are very variable in time and space. These variations cause more or less significant fluctuations in time and space of the solar radiation received on the ground. Two illustrations of the influence of clouds at different time and space scales are provided.

If it is relatively easy to calculate the solar radiation incident at the top of the atmosphere, the calculation is much more difficult for the radiation received on the ground, in various sky conditions with or without cloud. The radiation received on the ground depends on many variables, all of which vary in space and time. This chapter reviews the most influential. It highlights the complexity of calculating solar radiation at the ground level under changing sky conditions and presents a simplification of this calculation, which is of great practical significance.

6.1 Components of the solar radiation at ground level

6.1.1 Direct, diffuse, and reflected components – global radiation

Figure 6.1 is a schematic representation of the direct, diffuse, and reflected components of the radiation received by a tilted plane located on the ground, whose angle of inclination is β. In this graph, the solar zenithal angle is θ_S and the angle formed by the direction of the sun and the normal to the plane is θ.

The incident solar radiation at the top of the atmosphere is attenuated during the downward crossing in the atmosphere, in particular with multiple scattering, implying changes of direction of the photons compared to the original direction given by θ_S. The plane receives photons from various directions. The total or spectral radiation appearing to come from the direction of the sun is the direct component of the radiation. It is often represented by the letter B, which is the first letter of "beam". The total or spectral diffuse component gathers the photons coming from the other directions of the portion of the sky visible from the collector plane. This visible portion is the intersection between the sky and the hemisphere defined by the collector plane and its normal (Figure 6.1). This component is often noted D.

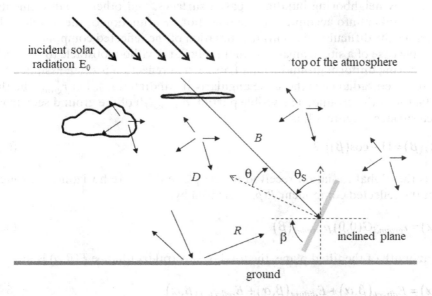

Figure 6.1 Schematic view of the direct (B), diffuse (D), and reflected (R) components of the radiation received by an inclined plane at ground whose inclination is β. θ_S is the solar zenithal angle, and θ is the angle formed by the direction of the sun and the normal to the plane.

Some authors use the term *beam component* to designate the direct component received at normal incidence. For my part, I note B_N the direct component at normal incidence and B this component in the general case. When the collector plane is horizontal, the following important relationship stands:

$$B = B_N \cos(\theta_S) \tag{6.1}$$

In grazing incidence, when the sun is on the horizon or below the horizon, θ_S is equal to or greater than $\pi/2$ and the direct component is extinguished. However, the solar rays still illuminate part of the sky vault that scatters them and the plane receives a nonzero diffuse component although the sun is no longer visible to an observer on the ground.

Figure 6.1 is related to Figures 4.11 and 4.12 (Chapter 4), which are schematic representations of the different paths of the solar rays incident at the top of the atmosphere and received by a horizontal plane located on the ground. The difference is that Figures 4.11 and 4.12 deal with the horizontal plane while the plane is tilted in Figure 6.1. In this case, a portion of the ground is visible from the collector plane. It is therefore necessary to add to the direct and diffuse components, a total or spectral reflected component noted R, due to the reflection of solar radiation by the ground.

The reflected component of the radiation strongly depends on the surroundings of the collector plane. For example, if this plane is in a valley, surrounded in certain directions by notable reliefs, there is reflection of the radiation on these reliefs. The reflected fluxes depend on the slopes and orientations of these reliefs and the nature of the soil. Another complex case is that of urban areas, where the solar rays can be reflected in multiple ways by neighboring buildings, paved surfaces, and other urban elements. It must also be taken into account that the reflection is a function of the wavelength, which increases the difficulty of accurate estimation of the reflected component.

In the simple case of a site in an area without relief and without obstacle to the horizon, a neighboring uniform ground with a Lambertian reflection is often assumed for which the reflected radiance is the same angularly in all directions. Let r_{ground} be the reflection factor of the ground. The visible portion $p_{ground}(\beta)$ of the ground seen from the inclined surface (Figure 6.1) is:

$$p_{ground}(\beta) = (1 - \cos(\beta))/2 \qquad (6.2)$$

If $G(0, 0)$ is the global radiation received by the plane if it were horizontal, a rough estimate of the reflected component $R(\beta, \alpha)$ is given by:

$$R(\beta,\alpha) = r_{ground} G(0,0) p_{ground}(\beta) \qquad (6.3)$$

If α is the azimuth of the tilted plane, the total or spectral irradiance $E(\beta, \alpha)$ is:

$$E(\beta,\alpha) = E_{direct}(\beta,\alpha) + E_{diffuse}(\beta,\alpha) + E_{reflected}(\beta,\alpha) \qquad (6.4)$$

where E_{direct}, $E_{diffuse}$ and $E_{reflected}$ are the three total or spectral components of the irradiance. The total or spectral irradiation $H(\beta, \alpha)$ obeys the same equation:

$$H(\beta,\alpha) = H_{direct}(\beta,\alpha) + H_{diffuse}(\beta,\alpha) + H_{reflected}(\beta,\alpha) \qquad (6.5)$$

where H_{direct}, $H_{diffuse}$ and $H_{reflected}$ are the components of the irradiation. If the collector plane is horizontal, it does not see the ground in the general case and the reflected component R is zero:

$$E(0,0) = E_{direct}(0,0) + E_{diffuse}(0,0) \qquad (6.6)$$

When it is necessary to clearly distinguish the radiation from its components, the radiation is called global radiation, often noted G, whether it is irradiance or irradiation. This sum is written generically as:

$$G = B + D + R \qquad (6.7)$$

Hence, G can be total or spectral global irradiance or total or spectral global irradiation.

6.1.2 Several definitions of the direct component

Several definitions of the direct component coexist which depend on the field of application, the measuring instruments, and other tools for estimating the direct component. This can create confusion and does not facilitate interdisciplinary exchanges.

Most of the numerical models simulating the radiative transfer in the atmosphere consider the sun as a point source. In this case, the total or spectral direct component is the radiance coming from this single point.

In a very different way, the total or spectral direct component of the radiation received on the ground may be defined as the sum of all the photons received on the flat collector in the solid angle under which the sun appears. Seen from the Earth, the solar disk has an angular diameter that varies slightly with the distance from the Earth to the sun and that is about half a degree ($0.53° \pm 0.008°$ of arc, or $32' \pm 0.5'$). This is the definition that I have adopted when introducing the direct component in Chapters 3 and 4.

The ISO standard[1] on terms in solar energy has a similar definition for the total radiation. However, it only mentions "a small solid angle centered on the solar disk" without any precision on the value of this angle. From a practical point of view, an instrument located on the ground, aiming with great precision at the center of the sun all the time and having an opening with an angular diameter of half a degree to receive only the flux coming from the direction of the solar disk, would be quite expensive to realize and also complex to operate, in particular to regularly check the alignment with the center of the sun. Current instruments measuring the total direct component follow the recommendations of the Guide to Meteorological Instruments and Methods of Observation[2] of the World Meteorological Organization. They aim the sun with fairly great precision and have a circular opening, with an angular diameter of 5°, which is ten times wider than that of the sun. It is this measured quantity that is called direct component in meteorological networks.

In fact, seen from the ground, the sun presents a halo due to the presence of scattering elements; the contribution of this halo to radiation is called the circumsolar contribution. The circumsolar contribution is formally part of the diffuse component. Practically because of the large opening of the instruments, it is measured, partially or totally depending on the case, with the direct component without it being easy to separate them.

The circumsolar contribution can have a more or less significant contribution to the measurement of the direct component. It is a function of several more or less known variables. In overcast conditions where the direct component is small, the contribution to the total irradiance is typically less than 10 W m^{-2} or less than 1 % in relative value. In the case of a very clear sky without cloud, the circumsolar contribution is of the same order than previously and accordingly rather negligible compared to the flux coming from the solar disk. The most critical cases are those with a high concentration of large size aerosols in the direction of the sun or with thin cirrus partially obscuring the sun. These large particles or ice crystals preferentially scatter the radiation forward, and the circumsolar contribution is then greater than in the case of scattering with smaller and fewer particles. It can represent up to 20 % of the measurement of the direct component.

In the case of spectral radiation, the dependency of the angular variation of the radiance of the solar disk with the wavelength must be taken into account. The smaller the wavelength, the greater the angular variation.

---

1 ISO 9488:1999. Solar Energy – Vocabulary. Confirmed in 2014, International Standard Organization, Geneva, Switzerland. [See in particular the definition of the direct component.]

2 Guide to Instruments and Methods of Observation. WMO-No. 8, 2018 edition, 2018, World Meteorological Organization, Geneva, Switzerland. [Vol 1 Measurement of Meteorological Variables, Chapter 7, Measurement of Radiation.]

Reader, I hope I have revealed to you the difficulty of estimating the direct component and the circumsolar contribution. I recommend reading the article by Blanc et al.[3] whose authors explore this complexity in detail. Among their conclusions, they recommend in particular to always clarify the definition adopted for the direct component, which I endeavor to do in this book.

6.1.3 Diffuse fraction – direct fraction

The diffuse fraction k_D of the radiation (irradiance, irradiation) is the ratio of the diffuse component D to the global radiation G received on a horizontal plane, provided G is not null:

$$k_D = D/G \tag{6.8}$$

The diffuse fraction is dimensionless. It quantifies the part of the diffuse radiation in the global radiation. It can be defined for a wavelength, a range of wavelengths, or the entire spectrum. By default, the latter prevails.

In principle, the diffuse fraction varies between 0 and 1. It is never 0 because even if the atmosphere is pure (no aerosol) and dry (no water vapor), the radiation is scattered by the air molecules and the diffuse component is nonzero.

Under cloud-free conditions and for the case of a horizontal plane, the diffuse fraction increases as the solar zenithal angle θ_S as the air mass increases. It reaches 1 when the direct component is zero, including when the sun is below the horizon. It increases as the number of diffusers increases, whether aerosols or air molecules, and therefore as the turbidity of the clear atmosphere increases. Very roughly, as a rule of thumb, the total diffuse fraction for a limpid sky is of the order of 0.3 in cloudless conditions. In other words, 30 % of the radiation comes from the sky vault, the rest being the direct component. If the clear sky is more turbid, due, for example, to suspended particles of desert origin, the diffuse fraction may be close to 0.5 or more.

As the optical depth of the cloud layer increases, the direct component decreases because the scattering is more intense than in clear sky and the diffuse fraction increases. The direct component is extinguished as the cloud optical depth is greater than about 3. In this case, the diffuse fraction is 1.

In the same way, the total or spectral direct fraction k_B can be defined as the ratio between the direct component and the global radiation. This quantity is less used.

6.2 Clearness index

6.2.1 Definition

The clearness index K_T is the ratio of the irradiance E or irradiation H received on a horizontal surface on the ground to the corresponding extraterrestrial irradiance E_0 or irradiation H_0, provided E_0 or H_0 is not null:

$$K_T = E/E_0 = H/H_0 \tag{6.9}$$

3 Blanc P., Espinar B., Geuder N., Gueymard C., Meyer R., Pitz-Paal R., Reinhardt B., Renne D., Sengupta M., Wald L., Wilbert S., 2014. Direct normal irradiance related definitions and applications: the circumsolar issue. *Solar Energy*, 110, 561–577, doi:10.1016/j.solener.2014.10.001.

The clearness index is dimensionless. The clearer the sky, the greater K_T. In principle, K_T varies between 0 and 1. There are special cases of multiple reflections on the clouds resulting in a momentary concentration of solar rays on the flat collector and therefore a clearness index greater than that expected for a cloudy sky. In these conditions, K_T may even be greater than 1.

What is the difference between the transmittance of the atmosphere and the clearness index? Both are ratios of irradiance or irradiation. The transmittance only quantifies the transmission through the atmosphere, while the clearness index quantifies the transmission and includes the reflection of the radiation by the ground. The clearness index is greater than the transmittance, except when the reflection by the ground is zero in which case the two are equal.

The clearness index can be total, i.e., calculated for the entire spectrum, or spectral, i.e., calculated at a given wavelength or for a specific spectral range, for example, $[\lambda_1, \lambda_2]$. Irradiances or irradiations, in the denominator and in the numerator, must correspond to the same spectral interval:

$$K_T[\lambda_1, \lambda_2] = E[\lambda_1, \lambda_2]/E_0[\lambda_1, \lambda_2] = H[\lambda_1, \lambda_2]/H_0[\lambda_1, \lambda_2] \tag{6.10}$$

The clearness index is usually calculated for a horizontal surface but can also be calculated for an inclined surface. Irradiances or irradiations, in the denominator and in the numerator, must correspond to the same receiving plane.

The clearness index can be defined for any duration, for example, 10 min, hourly, daily, monthly, or yearly. For example, the hourly clearness index $K_{T_{hour}}$ is the ratio of the hourly average of the irradiance E_{hour} or hourly irradiation H_{hour} to the corresponding extraterrestrial irradiance E_{0hour} or irradiation H_{0hour}:

$$K_{T_{hour}} = E_{hour}/E_{0hour} = H_{hour}/H_{0hour} \tag{6.11}$$

Pay attention to the calculation of the clearness index! As it is a ratio, it must be calculated as the ratio of the corresponding irradiances or irradiations, and not as the average of the ratios. Take the example of the daily clearness index. It is equal to the ratio of the daily averages of irradiance, or of the daily irradiations, and not to the mean of the hourly clearness indices of the day. Similarly, the monthly clearness index is the ratio of the monthly averages of irradiance, or of the monthly irradiations, and not the monthly mean of daily clearness indices of the month.

Whether for a spectral interval or total spectrum, and provided E_0 or H_0 is not null, one can define the direct clearness index $K_{T_{direct}}$ for the direct component E_{direct}, or H_{direct}, and the diffuse clearness index $K_{T_{diffuse}}$ for the diffuse component $E_{diffuse}$, or $H_{diffuse}$:

$$K_{T_{direct}} = E_{direct}/E_0 = H_{direct}/H_0 \tag{6.12}$$

$$K_{T_{diffuse}} = E_{diffuse}/E_0 = H_{diffuse}/H_0 \tag{6.13}$$

If $E_{Ndirect}$ and E_{0N} are respectively the direct component received on a plane normal to the direction of the sun at ground level and the extraterrestrial irradiance at normal incidence, the direct clearness index $K_{T_{Ndirect}}$ at normal incidence is defined as:

$$K_{T_{Ndirect}} = E_{Ndirect}/E_{0N} \tag{6.14}$$

Since as seen a few pages before in equation (6.1), the following relationship holds between E_{direct} and the direct component $E_{Ndirect}$ received on the ground at normal incidence:

$$E_{direct} = E_{Ndirect} \cos(\theta_S) \tag{6.15}$$

as well as at the top of the atmosphere:

$$E_0 = E_{0N} \cos(\theta_S) \tag{6.16}$$

it comes:

$$K_{T_{Ndirect}} = E_{Ndirect}/E_{0N} = \left[E_{direct}/\cos(\theta_S) \right]/\left[E_0/\cos(\theta_S) \right] = K_{T_{direct}} \tag{6.17}$$

$K_{T_{Ndirect}}$ is equal to the direct clearness index $K_{T_{direct}}$.

The clearness index has the advantage of eliminating the influence of the variation in extraterrestrial irradiance, due to the variation of the distance between the sun and the Earth, of the declination and of the solar zenithal angle, or of the angle of incidence in the case of an inclined plane. Consequently, it has the advantage of eliminating the influences of the seasons and latitudes due to these variations in E_0, or H_0.

6.2.2 An example of 15-min clearness indices

Let me illustrate this advantage to compare the types of sky for the same site between the different seasons. I chose a marine site in the North Atlantic Ocean to avoid the reflection effects. Latitude is 50° and longitude is −20°. I randomly selected 4 days in 2005: 01-10 (10 January), 04-10 (10 April), 07-10 (10 July), and 10-10 (10 October). I used here estimates made by models and not measurements made by instruments on the ground. This does not matter since the object is to illustrate the interest of the clearness index.

I have plotted in Figure 6.2 the profile of the irradiance received on the surface on a horizontal plane averaged over 15 min, for each of the 4 days. I drew three profiles of 15-min irradiances: the actual one, the one assuming a cloudless sky (clear sky), and the extraterrestrial one. Extraterrestrial irradiance is greatest in July, then in April, then in October, and it is minimal in January. It is the same hierarchy for clear-sky irradiance and actual irradiance in general. There are many clouds on 10 January, and the actual profile of irradiance shows low values compared to that of the clear sky. The curve is quite smooth, which indicates a fairly constant cloud cover throughout the day in time. For the other 3 days, the actual irradiance is fairly close to that of clear sky with marked breaks denoting an alternation of clear sky and cloud passages.

In Figure 6.2, the solar irradiation received at the top of the atmosphere is not identical between the 4 days. As a consequence, it is difficult to compare the influence of the atmosphere during these days. However, this can be done by looking at the clearness index as shown in Figure 6.3, which exhibits the profiles of K_T for the cases of actual and clear-sky conditions.

The curves are less sharp around noon in true solar time compared to Figure 6.2. The curves of the clear-sky clearness index are similar for the 3 days in April, July,

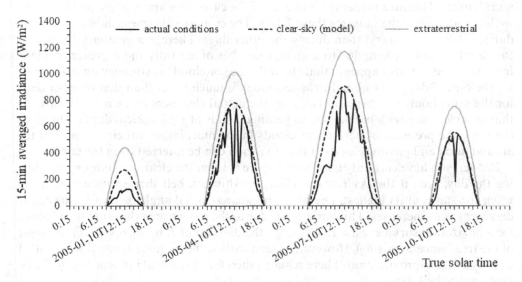

Figure 6.2 Daily profiles of the 15-min average of total irradiance received on a horizontal plane at ground level, for the site of latitude 50° and longitude −20°, for 4 days in 2005: 01-10, 04-10, 07-10, and 10-10. Three profiles of 15-min irradiances are plotted: the actual one, the one assuming a cloudless sky (clear sky), and the extraterrestrial one. (Source of data: HelioClim-3 database and McClear clear-sky model, available at SoDa Service (soda-pro.com).)

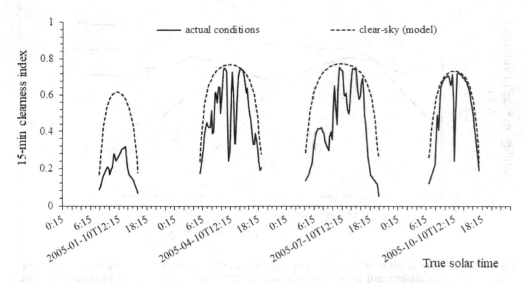

Figure 6.3 Daily profiles of the 15-min clearness index for the site of latitude 50° and longitude −20°, for 4 days in 2005: 01-10, 04-10, 07-10 and 10-10. Two profiles of clearness index are plotted: the actual one and the one assuming a cloudless sky (clear sky). (Source of data: HelioClim-3 database and McClear clear-sky model, available at SoDa Service (soda-pro.com).)

and October. This means that the influence of the cloud-free atmosphere on the downwelling radiation is the same for these 3 days. The clear-sky clearness indices are lower during 01-10 (10 January) than during the other days. There is a greater influence of the cloudless atmosphere due to a greater number of aerosols and a greater content in water vapor. It also appears that the influence of clouds is stronger on 01-10 than for the other 3 days: The actual clearness index is much lower than that of other days for the same hour. In April and October, the actual clearness index is fairly close to that of a clear sky with breaks due to passing clouds of high optical depth. On 07-10 (10 July), the presence of permanent clouds in the morning and their disappearance around noon and passing clouds in the afternoon can be inferred from the curves.

Reader, you have certainly noticed in Figure 6.3 that the clearness index varies during the day, even if the sky remains clear. It exhibits a bell shape with a maximum when the sun is at its highest, i.e., when the solar zenithal angle θ_S is at its lowest. It decreases as θ_S increases. This variation is due to the influence of the air mass m. Since for a horizontal surface, $E_0 = E_{0N} \cos(\theta_S)$, the division of E by E_0 involves a division of the irradiance by $\cos(\theta_S)$. However, as seen earlier, the air mass is equal to $1/\cos(\theta_S)$ only as a first approximation. There remains therefore a residual influence of θ_S which explains the bell shape.

6.2.3 An example of monthly clearness indices

I offer another illustration of the value of the clearness index, but this time I am dealing with monthly indices. I selected three points of longitude 0° and latitudes 0°, 30°, and 60°, respectively. Figure 6.4 exhibits the yearly profile of monthly averages of

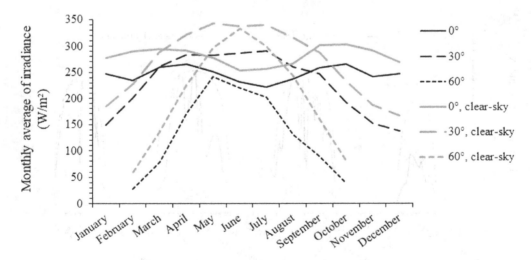

Figure 6.4 Yearly profile of monthly averages of total irradiance received on a horizontal plane at ground level during the year 2006 at three sites of longitude 0° and latitudes 0°, 30°, and 60°, respectively. Two profiles are plotted: the actual one and the one assuming a cloudless sky (clear sky). There are not enough data during January, November, and December at 60° to obtain a reliable average. (Source of data: HelioClim-3 database and McClear clear-sky model, available at SoDa Service (soda-pro.com).)

irradiance received on a horizontal plane at ground level during the year 2006. As for the previous example, I give the irradiances in the assumption of a clear sky and for the actual sky conditions because discussing these two cases allows to better understand on the one hand, the clearness index and, on the other hand, the predominant influence of clouds.

In both cases, there is a large variation in irradiance over the course of the year which increases as one moves away from the equator (Figure 6.4). The shape of the profiles of the irradiance for the clear sky is quite close to what was seen for the irradiance received by a horizontal plane at the top of the atmosphere in Chapter 3. The curves for the actual conditions are lower than those for the clear-sky conditions as expected. It is difficult to simply compare the curves between the three latitudes with regard to the optical quality of the atmosphere. Does the atmosphere attenuate the solar radiation more at one of these latitudes? On the contrary, is the atmospheric attenuation the same whatever the latitude?

The clearness index is one element to quickly answer these questions. Figure 6.5 exhibits the yearly profile of the corresponding monthly clearness indices. For clear-sky conditions, K_T is greater at the point of latitude 30° than at the other two: The clear sky is more limpid on average. K_T is greater at the site of latitude 60° than at equator from March to September and lower in the other months. In general, the monthly clearness index in clear-sky conditions tends to remain constant over the year at the three sites and offers little variation compared to the irradiance.

The situation is different in the actual conditions. The clearness indices are less than those under clear skies, which indicates that there were cloudy situations during each month. The clearness indices at 0° and 30° are quite similar between May and January:

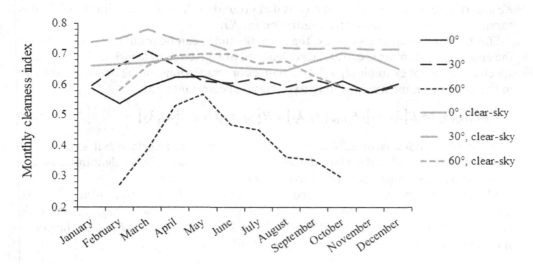

Figure 6.5 Yearly profile of monthly clearness index during the year 2006 at three sites of longitude 0° and latitudes 0°, 30°, and 60°, respectively. Two profiles are plotted: the actual one and the one assuming a cloudless sky (clear sky). There are not enough data during January, November, and December at 60° to obtain a reliable clearness index. (Source of data: HelioClim-3 database and McClear clear-sky model, available at SoDa Service (soda-pro.com).)

The properties of the clouds are similar on average. The attenuation of the radiation by clouds is more marked at $0°$ than at $30°$ from February to April for the year 2006. At $60°$, the clearest month is June with $K_T = 0.57$, which is however much lower than the clearness index for the clear sky of the same month: 0.70. Clouds are often present throughout the year at this site at $60°$, and the clearness index is often low with a peak in June.

As a summary, the clearness index quantifies the attenuation of radiation by the atmosphere in a simple way, for daily, monthly, or yearly periods. If the clear sky is limpid, the relative extinction of the radiation is about 20 %; the daily, monthly, or annual clearness index is approximately 0.8. A clearness index less than 0.2 indicates an overcast sky with fairly optically thick clouds. If $0.3 \leq K_T \leq 0.6$, the cloud cover has a low optical depth, or may consist of fragmented clouds having fairly large optical depths.

6.3 Clear-sky index

The clear-sky index Kc is the ratio of the irradiance E, or irradiation H, received at ground to the irradiance E_{clear}, or irradiation H_{clear}, received at ground in clear-sky conditions, provided E_{clear} or H_{clear} is not null:

$$Kc = E/E_{clear} = H/H_{clear} \tag{6.18}$$

The clear-sky index is sometimes called cloud modification factor. It is dimensionless. Kc varies between 0 and 1. The greater the attenuation of the radiation by the clouds, the smaller Kc. In principle, Kc is 1 if the sky is clear. In practice, E_{clear} or H_{clear} is only known using a clear-sky model and it is possible that the estimate of E_{clear} or H_{clear} is less than the actual E_{clear} or H_{clear}, in which case Kc will be greater than 1.

What is the difference between Kc and the clearness index K_T? Normalization for Kc is carried out by the radiation received at ground in cloudless conditions, while it is carried out by the extraterrestrial radiation for K_T.

Similarly to the clearness index, the clear-sky index can be total, i.e., calculated for the entire spectrum, or spectral, i.e., calculated at a given wavelength or for a specific spectral range, for example $[\lambda_1, \lambda_2]$. Irradiances or irradiations, in the denominator and in the numerator, must correspond to the same spectral interval:

$$Kc[\lambda_1, \lambda_2] = E[\lambda_1, \lambda_2]/E_{clear}[\lambda_1, \lambda_2] = H[\lambda_1, \lambda_2]/H_{clear}[\lambda_1, \lambda_2] \tag{6.19}$$

The clear-sky index is usually calculated for a horizontal surface but can also be calculated for an inclined surface. Irradiances or irradiations, in the denominator and in the numerator, must correspond to the same receiving plane.

The clear-sky index can be defined for any duration, for example, 10 min, hourly, daily, monthly, or yearly. For example, the hourly clear-sky index Kc_{hour} is the ratio of the hourly average of the irradiance E_{hour} or hourly irradiation H_{hour} to the corresponding clear-sky irradiance E_{clear_hour} or irradiation H_{clear_hour}:

$$Kc_{hour} = E_{hour}/E_{clear\_hour} = H_{hour}/H_{clear\_hour} \tag{6.20}$$

Please reader, pay attention to the calculation of the clear-sky index! As it is a ratio, it must be calculated as the ratio of the corresponding irradiances or irradiations, and not as the average of the ratios. Take the example of the daily clear-sky index. It is equal to the ratio of daily averages of irradiance, or daily irradiations, and is not to the mean of the hourly clear-sky indices of the day. Similarly, the monthly clear-sky index is the ratio of monthly averages of irradiance, or monthly irradiations, and is not the monthly mean of daily clear-sky indices of the month.

By way of illustration, I represent in Figure 6.6 the clear-sky indices Kc at the same point in the North Atlantic Ocean, of latitude 50° and longitude −20° for the same 4 days in Figures 6.2 and 6.3. By definition, the clear-sky index in cloudless conditions is equal to 1 whatever the solar zenithal angle (dashed curve). Changes in actual clear-sky index are due to the presence of clouds and their fluctuating properties as discussed in Figure 6.3. Note in particular the weak clear-sky indices during 01-10 for which the influence of clouds is very strong. In October (10-10), the actual clear-sky index is fairly close to that of a clear sky with a few breaks due to passing clouds of high optical depth. In April and July, a cloud deck is present in the morning and at the end of the day; in the middle of the day, marked breaks are observed in actual profiles denoting an alternation of clear sky and cloud passages.

Whether for a spectral interval or total spectrum, the diffuse clear-sky index $Kc_{diffuse}$ can be defined for the diffuse component $E_{diffuse}$, or $H_{diffuse}$:

$$Kc_{diffuse} = E_{diffuse}/E_{diffuse\_clear} = H_{diffuse}/H_{diffuse\_clear} \qquad (6.21)$$

where $E_{diffuse\_clear}$ and $H_{diffuse\_clear}$ are respectively the diffuse components of the irradiance and irradiation in cloud-free conditions and are not null.

In the same way, the spectral or total direct clear-sky index Kc_{direct} can be defined for the direct component E_{direct}, or H_{direct}:

$$Kc_{direct} = E_{direct}/E_{direct\_clear} = H_{direct}/H_{direct\_clear} \qquad (6.22)$$

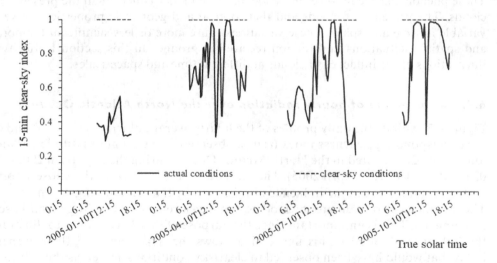

Figure 6.6 Daily profiles of the 15-min clear-sky index at the site of latitude 50° and longitude −20°, for 4 days in 2005: 01-10, 04-10, 07-10, and 10-10. The clear-sky index in cloudless (clear-sky) conditions is equal to 1 by definition. (Source of data: HelioClim-3 database and McClear clear-sky model, available at SoDa Service (soda-pro.com).)

where E_{direct_clear} and H_{direct_clear} are respectively the direct components of the irradiance and irradiation in cloud-free conditions and are not null.

Similarly to what was done for the clearness indices KT_{direct} and $KT_{Ndirect}$, it can be demonstrated that the clear-sky index Kc_{direct} for a horizontal plane is the same than the clear-sky index $Kc_{Ndirect}$ for a plane normal to the direction of the sun:

$$Kc_{direct} = Kc_{Ndirect} \tag{6.23}$$

As already mentioned previously, the irradiance E_{clear} received on the ground that would be measured if the sky were cloudless cannot be measured at the same time as the irradiance E if the actual situation is cloudy because it would imply that the sky is both cloudless and cloudy. Therefore, the clear-sky index is an index based on a modeling of the irradiance E_{clear} since the latter is not accurately known. Consequently, there may be some confusion as to the use of this index because unlike the clearness index, it is not standardized by an easily calculable quantity with sufficient accuracy and accepted by the different communities. Clear-sky indices and the results obtained from their exploitation depend on the clear-sky model used. This model should therefore be clearly mentioned. However, this index is a very good computational intermediary while having a physical meaning that is easy to grasp.

6.4 The prominent role of clouds

As seen previously, the attenuation of solar radiation, total or spectral, by the clouds is preponderant in front of the other attenuators. The clouds induce a more or less marked darkening of the sky for the observer on the ground. Dust storms, or ash clouds created by a volcanic eruption, can create complete darkness quite suddenly. These phenomena are however more localized and less frequent than the presence of clouds. The presence of clouds and their optical and geometric properties are very variable in time and space. These variations cause more or less significant temporal and spatial fluctuations in radiation received at ground. In this section I offer two illustrations of the influence of clouds at different time and space scales.

6.4.1 An example of hourly radiation over the North Atlantic Ocean

Figure 6.7 exhibits the daily profiles of the hourly average of irradiance (left) and of the corresponding clearness index (right), observed at the point of latitude 50° and longitude −20°, located in the North Atlantic Ocean, during the days from 2005-04-02 to 2005-04-05 (2–5 April 2005). The receiving plane is horizontal. I chose a point on the ocean to reduce the influence of reflection by the surroundings of the plane. The irradiances are estimates by models, not measurements made by ground-based instruments. This is unimportant since the purpose of this discussion is to illustrate the influence of clouds. This figure also shows the irradiance and the clearness index that would have been observed in clear-sky conditions as well as the extraterrestrial irradiance.

The irradiance profiles exhibit a bell shape in the case of clear sky as for the top of the atmosphere. The irradiance is greater when the solar zenithal angle is small and decreases when this angle increases.

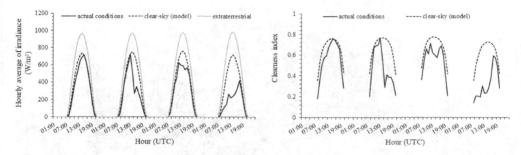

Figure 6.7 Daily profiles of the hourly average of total irradiance at ground level (a) and the corresponding clearness index (b). The latitude and longitude of the site are 50° and −20°, respectively. The profiles are drawn from 2 to 5 April 2005, for the actual conditions and assuming a cloudless sky (clear sky). The extraterrestrial irradiance is also plotted on the left. (Source of data: HelioClim-3 database and McClear clear-sky model, available at SoDa Service (soda-pro.com).)

There is little variation in the distance between the sun and the Earth and in declination during these 4 days. Consequently, the profiles of the extraterrestrial irradiance (gray curves) are the same for the 4 days. The maxima are reached for the time stamp 14:00 UTC, end of the integration time, because of the longitude offset of 20° to the west. The profiles of irradiance in clear-sky conditions are very similar between the first 3 days, with maxima of 736, 750 and 761 W m^{-2}, respectively. The clearness indices (Figure 6.7b) are close to or greater than 0.7 for a large part of each of these days. During the fourth day, the aerosol load increased sharply, mainly sulfates and desert dust. The irradiance in clear sky is then weaker than for the previous days; the maximum is 717 W m^{-2} and the clearness indices are less than 0.7.

On 04-02 (2 April), the actual irradiance is close to that of clear sky: The black curve is superimposed on the dotted one. The following day, on 04-03, the actual irradiance is close to that of clear sky in the morning then decreases sharply due to the presence of clouds. During this afternoon, the clearness index is less than 0.4, that is to say, that 40 % of the irradiance received at the top of the atmosphere is received by the horizontal plane on the ground. The sky is clear on 04-04 (4 April), except around solar noon (14:00 UTC), when the clouds strongly attenuate the solar radiation. The following day (04-05) is very cloudy with a clearness index less than 0.4, which increases up to 0.6 in the middle of the afternoon.

Beyond this illustration, I would like you, reader, to note that on the one hand, the attenuation due to clouds is much stronger than that due to an increase in the aerosol load, and on the other hand, the temporal variations of the clouds properties cause strong temporal variations on the irradiance received on the ground.

6.4.2 Map of a multi-year average of the total solar irradiance at ground

I have already shown that at the top of the atmosphere, the annual average of irradiance on a horizontal plane has a fully latitudinal geographic distribution, independent of longitude, which is recalled here in Figure 6.8. The distribution is symmetrical with

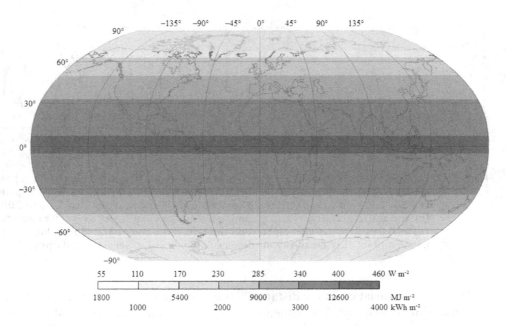

Figure 6.8 Map of the annual average of the extraterrestrial total irradiance received on a horizontal plane and the corresponding irradiation. Latitudes are shown on the left and longitudes at the top.

respect to the equator: The values are the same in the two hemispheres at equivalent latitudes. The irradiance is greatest at the equator and decreases from the equator to the poles. A latitudinal distribution is also called a zonal distribution.

Incidentally, I recall the conversion between the annual average irradiance (W m^{-2}) and the annual irradiation, in MJ m^{-2} or Wh m^{-2}. If the irradiance is equal to 1 W m^{-2}, the annual irradiation is computed by:

$$(1 \text{ W m}^{-2}) \times (365 \text{ day}) \times (86{,}400 \text{ s day}^{-1}), \text{ i.e. } 31.536 \text{ MJ m}^{-2}$$
$$\text{or } (1 \text{ W m}^{-2}) \times (365 \text{ day}) \times (24 \text{ h day}^{-1}), \text{ i.e. } 8.76 \text{ kWh m}^{-2}$$

The presence of clouds in the atmosphere creates two effects: a decrease in irradiance and a non-zonal geographic distribution. This is also true in the absence of clouds, but with much less intensity. In order to illustrate the effects of clouds, Figure 6.9 presents a map of the annual average irradiance received on a horizontal plane at ground for the period 1990–2004. First of all, a decrease of the values is observed compared to Figure 6.8. Attenuation comes mainly from the presence of clouds. One may retain that a rough estimate of the annual average of the clearness index is around 0.45 at mid-latitudes.

Note also that the atmospheric extinction is not the same everywhere, because of the geographic heterogeneity of the cloud cover and its effects on solar radiation. The distribution is fairly latitudinal between the poles and latitudes −40° and +40°, respectively, although several indentations are visible. The geographic distribution is more complex between latitudes −40° and 40°.

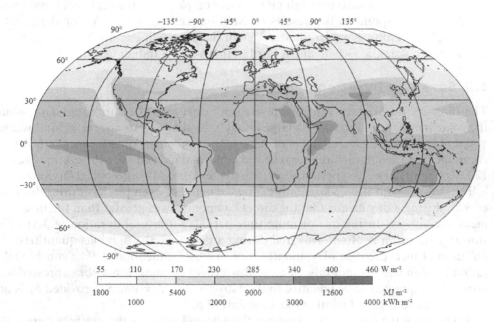

Figure 6.9 Multi-annual (1990–2004) average of the solar total irradiance received on a horizontal plane on the ground and the corresponding irradiation. Latitudes are shown on the left and longitudes at the top.

The impact of large mountain ranges and large plateaus on the irradiance may be observed in Figure 6.9 because they have an influence on the presence of clouds. In addition, with their altitude being high, the atmosphere layer above the mountains is thinner than above the ocean and the atmospheric attenuation is less. There is a kind of barrier along the west coast of America that can be attributed to the presence of the Rocky Mountains in the north and the Andes Cordillera in the south. A similar structure is observed due to the Australian Cordillera along the east coast of Australia. The set of mountain ranges in Northeast Africa – Tibesti, Marrah Mountains, and the Ennedi Plateau – corresponds to a greater irradiance compared to the neighboring regions. The same goes for the Brazilian Plateau, or the Himalayas.

The influence of large ocean currents on the presence of clouds is also visible and is marked by indentations in relation to the latitudinal distribution. One can thus recognize in the Pacific Ocean, the cold Current of California, which flows from north to south along the west coast of North America, or that of Oyashio along the northeast coasts of Asia and Kamchatka peninsula, or that of Humboldt, also known as Current of Peru, which flows north along the west coast of South America. In the North Atlantic Ocean, the influence of the warm Gulf Stream flowing northeast along the west coast of America and that of the cold Labrador Current can be seen. Note in the South Atlantic Ocean the indentations due to the warm Current of Brazil flowing toward the south, along the west coast of America, and the cold Current of Benguela along the west coast of the Africa, as well as the anticyclonic circulation loop formed by these two currents and the South Equatorial and South Atlantic Currents. The influence of the warm Agulhas Current along the east coast of Africa is also visible.

On land, in addition to the high mountains and plateaus, the regions of maximum irradiance correspond to the deserts of Namibia and Australia. As for the oceans, three regions exhibit great irradiance: one in the Southern Pacific Ocean, another in the Southern Atlantic Ocean, and a third in the Western Indian Ocean.

6.5 Sunshine duration

The sunshine duration or sunshine hours is a measure of the duration during which the solar irradiance at ground is large enough for shadows to appear behind illuminated objects. More precisely, it is defined as the sum of the time for which the direct component of the solar irradiance received at normal incidence on the ground exceeds 120 W m^{-2} during a given period according to the World Meteorological Organization.

The sunshine duration S is usually expressed in h. For example, a sunshine duration of 3.5 h during a day means that the direct component was greater than the threshold mentioned above during one or more time periods for a total duration of 3.5 h. The sunshine duration is often measured during a day. Like the previous quantities, the duration of measurement (day, month, year...) must be mentioned. The sunshine duration is often used as monthly, or annual, averages or sums. It may be expressed as a normalized quantity (S/S_0) relative to the astronomical daytime S_0 provided S_0 is not null. This ratio is called relative sunshine duration.

At the top of the atmosphere, whatever the date and location, the sunshine duration S is equal to the daytime S_0 and the ratio S/S_0 is equal to 1. On the ground, the sunshine duration S varies as a function of time, geographic location, and meteorological conditions. If the sky is overcast during the day, the clouds obscure the sun all day, the sunshine duration is zero, and the relative sunshine duration is also zero. In clear-sky conditions, S mainly depends on the water vapor content and the aerosol load. The cleaner the atmosphere, the longer the sunshine duration, and the greater the relative sunshine duration.

A non-leap year includes 8760 h. Table 2.1 reports that the annual average of the daytime is around 12 h. Thus, the annual sum of sunshine duration at a given location cannot exceed approximately 4380 h. The annual sum of sunshine duration observed in the world varies between approximately 1000 h in the polar zones and the very cloudy zones and 4000 h in the zones where the cloudiness is small all year round.

6.6 The most important variables for the solar radiation at ground

While it is relatively easy to calculate the extraterrestrial radiation, i.e., the incident solar radiation at the top of the atmosphere, the preceding discussions and illustrations show that the calculation is more difficult for the radiation received on the ground, whether in clear-sky conditions or cloudy conditions. It is not so much that models or computer resources are lacking; it is above all accurate and localized data that are lacking. Indeed, the radiation received on the ground depends on many variables, all of which vary in space and time.

Let me review the variables most influencing the radiation received on a horizontal plane at ground. First, the date and the geographic coordinates are used to calculate the solar zenithal angle θ_S as well as the radiation received at the top of the atmosphere. If a particular spectral interval is of interest, a spectral distribution of extraterrestrial radiation must be known.

The elevation of the ground above the mean sea level is a variable which is fairly easy to know nowadays and which is invariant, except in special cases. The higher the elevation, the thinner the atmospheric layer above the site considered, and the lower the atmospheric attenuation.

In clear-sky conditions, the variable θ_S still plays an important role since it determines the air mass for each absorbing or scattering element. The concentrations of absorbing gases and the spectral distribution of the absorption coefficients must be known for the selected date and location. For the evaluation of total radiation, it is generally sufficient to consider the ozone and water vapor contents only, which are very variable, and to prescribe the concentrations of the other gases to standard values. As absorption depends on local conditions of temperature, density, and pressure, the vertical profile of these variables, as well as that of the volume mixing ratio of absorbing gases excluding ozone and water vapor, must be known. This profile also allows the calculation of the scattering effects of air molecules. Finally, the optical properties of aerosols must be known, which are highly variable. Usually, classes of aerosols are used, for example, desert dust, or soot, which have been assigned average optical properties. The aerosol load is given by an optical depth, known at one or more wavelengths, for example, 550 and 1020 nm, or else the optical depth at a given wavelength and Ångström exponent.

Cloud properties are highly variable and very difficult to know in detail at any time. Regarding the radiation received on the ground, the most important among them is by far the optical depth. The greater the optical depth, the greater the attenuation by the cloud, and the weaker the influences of the other cloud properties. The phase of the cloud must be determined if possible: Is it composed of water droplets, ice crystals, or a mixture of the two? This determination is quite often coupled with that of the altitude of the top of the cloud, using the vertical profile of temperature and density. If the summit altitude is low, then it is water droplets. On the contrary, if it is high, then it could be ice crystals. But if in addition, the optical depth is high, then the cloud could be of the cumulonimbus type and include water droplets in the lower part and ice crystals in the upper part. Other variables are involved, such as the liquid or solid water content, or the effective diameter of the droplets or crystals which has an influence on the scattering pattern. These variables are often unknown, and default values are often adopted.

Finally, in the case of clear sky as well as cloudy sky, variables are needed to describe the influence of the reflection of the ground on total or spectral radiation. As seen earlier, this influence of the ground itself depends on the radiation received, in particular on the diffuse fraction. This fraction strongly depends on the sky conditions; it is around 0.3 for a limpid sky and 1 for optically thick clouds. If the nature of the ground is known, published tables can be used that give a correspondence between the nature of the ground and the albedo. It is best to use bidirectional reflectance distribution functions.

At this point in the presentation, reader, you may be frightened by the number of variables that must be known at any time and any place, and then implemented at every moment to calculate the radiation received on the ground. You can because, indeed, the task is difficult. This is why at the time of this writing, the solar radiation received on the ground is not a meteorological variable known with exactitude at any place and any time.

One of the practical consequences for this work is that it is not as easy to give accurate values of the radiation received on the ground as I did for the solar radiation incident at the top of the atmosphere; there are many local variables to consider.

In order to obtain very approximate values of the irradiance on the ground, in particular around the solar noon, or on daily, monthly, or annual average, I can suggest to you, reader, to use some empirical rules linking the clearness index and the atmospheric attenuation. $K_T \geq 0.7$ indicates clear sky, and $K_T \leq 0.2$ indicates overcast sky with fairly optically thick clouds. The average of K_T observed at mid latitudes is around 0.45. The irradiance on the ground is given by multiplying the extraterrestrial irradiance, given in Chapter 3, by the chosen clearness index.

6.7 Decoupling the effects of clear atmosphere and clouds

A few lines above, I wrote on the difficulty of calculating the radiation received on the ground in all sky conditions, due in particular to the lack of measurements of all variables involved in this calculation. To this must be added the practical difficulties of calculation in terms of computer resources and data transmission. This section proposes an approximation that can be adopted by models for fast calculations of the radiation. It concerns both total radiation and spectral radiation.

It has been observed that, except in special cases of concentration of solar rays on the sides of the clouds, the irradiance E_{clear} received on the ground under clear-sky conditions is greater than that received under cloudy conditions. This amounts to saying that E_{clear} is an upper bound for irradiance E received on the ground in all sky conditions and that the effects of clouds amount to attenuating E_{clear}. This observation can be used to simplify the calculation of E by first calculating E_{clear} and then by calculating the attenuation of E_{clear} by the clouds.

The approach uses the clear-sky index Kc, also known as the cloud modification factor, and the direct clear-sky index Kc_{direct}, seen earlier in this chapter. It consists in using a clear-sky model to calculate E_{clear} and its direct component B_{clear}, and then in calculating Kc and Kc_{direct} according to the optical properties of the clouds to obtain the global irradiance E and its direct component B. The diffuse component D is then calculated by subtracting B from E.

However, formally, the indices Kc and Kc_{direct} depend on the constituents of the clear atmosphere and are expected to change with clear-atmosphere properties since the clouds and other atmospheric constituents are mixed up in the atmosphere. Consequently, they should be calculated taking into account also these constituents and not only the properties of clouds. Let P_{clear} be the set of variables describing the constituents of the clear-atmosphere input to the clear-sky model. For example, P_{clear} can be made up of the following:

- the ozone and water vapor contents in the atmospheric column,
- the vertical profiles of temperature, density, pressure, and volume mixing ratio of absorbing gases, excluding ozone and water vapor,
- the optical properties of aerosols, such as the aerosol type and their optical depth at several wavelengths.

In a similar way, let P_{cloud} be the set of variables describing the clouds used by the corresponding model. For example, P_{cloud} can be formed by the following:

- the cloud optical depth,
- the cloud phase,
- the liquid water content,
- the effective diameter of the water droplets and ice crystals,
- the vertical position of the cloud.

The global irradiance E is calculated by taking into account all these variables a priori:

$$E = E_{clear}\left(E_0, z, \theta_S, \rho, P_{clear}\right) Kc\left(z, \theta_S, \rho, P_{clear}, P_{cloud}\right) \qquad (6.24)$$

where E_0 is the extraterrestrial irradiance, z is the elevation of the ground above the mean sea level, θ_S is the solar zenithal angle, and ρ is the bidirectional reflectance distribution function. Likewise, the direct component B of the irradiance is calculated a priori as follows:

$$B = B_{clear}\left(E_0, z, \theta_S, \rho, P_{clear}\right) Kc_{direct}\left(z, \theta_S, \rho, P_{clear}, P_{cloud}\right) \qquad (6.25)$$

Reducing the number of variables to calculate Kc or Kc_{direct} simplifies the calculations and increases their speed. This is why several authors, notably Oumbé et al. (2014),[4] investigated numerically the influence of variations in P_{clear} on these indices by systematically scanning the domains of variation of each of the variables listed above. The published results show that the influence of variations in P_{clear} on Kc or Kc_{direct} can be small, by a few percent in relative value, and can be neglected as a first approximation. The error made is similar to that affecting the radiation measurements made by pyranometers.

Some cases exist where the relative error made by neglecting the influence of P_{clear} on Kc or Kc_{direct} becomes more noticeable. When the ground is very reflective, like fresh snow, for example, the contribution to E of the multiple reflections and backscatterings between the ground and the clouds increases compared to a poorly reflecting ground and the influence of the variations in P_{clear} on Kc or Kc_{direct} is higher. This influence increases with the optical depth of the cloud. When the solar zenithal angle θ_S becomes large, typically greater than $60°$, the optical path of the rays increases sharply and the influence of variations in P_{clear} on Kc or Kc_{direct} is higher. However, it should be noted that under such conditions, the irradiance E is low, and that the error is also small in absolute value.

What to conclude from it? Saying that the influence of the variations in P_{clear} on Kc or Kc_{direct} is weak amounts to saying that the first derivatives $\partial Kc/\partial P_{clear}$ and $\partial Kc_{direct}/\partial P_{clear}$ are zero as a first approximation. It follows that one can write:

$$E \approx E_{clear}\left(E_0, z, \theta_S, \rho, P_{clear}\right) Kc\left(z, \theta_S, \rho, P_{clear\_type}, P_{cloud}\right) \qquad (6.26)$$

4 Oumbé A., Qu Z., Blanc P., Lefèvre M., Wald L., Cros S., 2014. Decoupling the effects of clear atmosphere and clouds to simplify calculations of the broadband solar irradiance at ground level. *Geoscientific Model Development*, 7, 1661–1669, doi:10.5194/gmd-7-1661-2014.

$$B \approx B_{clear}\left(E_0, z, \theta_S, \rho, P_{clear}\right) Kc_{direct}\left(z, \theta_S, \rho, P_{clear\_type}, P_{cloud}\right) \tag{6.27}$$

where P_{clear_type} is a standard or average vector of variables describing the clear atmosphere, chosen arbitrarily from the P_{clear} set. Thus, with each computation of Kc or Kc_{direct}, one does not use the vector P_{clear} used for the computation of E_{clear} or B_{clear} but the vector P_{clear_type}. This thus decreases the number of variables for the calculation of Kc or Kc_{direct} since the variables describing the clear atmosphere are fixed. Several choices are possible for the vector P_{clear_type} as long as the values chosen are not extreme. For example, the aforementioned authors propose to set the elevation z to 0 and to use the following vector P_{clear_type}:

- an atmospheric ozone content of 300 Dobson units,
- an atmospheric column content of water vapor of 35 kg m^{-2},
- the profiles called *midlatitude summer* from the set of profiles offered by the Air Force Geophysics Laboratory (AFGL) in the United States, for the vertical profiles of temperature, density, pressure, and volume mixing ratio of absorbing gases, excluding ozone and water vapor,
- an optical depth of 0.2 at 550 nm for aerosols as well as an Ångström exponent of 1.3.

What are the practical benefits? Thanks to this approach, the radiation calculations can be decoupled. On the one hand, a clear-sky model can be used to estimate the radiation under clear-sky conditions E_{clear} and B_{clear}, and on the other hand, another model can be used to estimate the attenuation due to the clouds Kc and Kc_{direct}. Each model has its own set of inputs. The models are often in the form of one abacus or several abaci (also known as look-up tables) to save computing time. The less input the abaci have, the less space they take up in computer memory, and the faster the calculations. Having two separate models makes it possible to have two sets of abaci which each have fewer inputs than an abacus intended to calculate the radiation under all sky conditions. Reduced abaci are also easier to handle and store and faster to recalculate if one of the models is modified.

Chapter 7

Spectral distribution of the solar radiation at ground

At a glance

The solar radiation received at ground level has a spectral distribution quite similar to that received at the top of the atmosphere, but it is lower at all wavelengths. The shape of the spectral distribution of the radiation varies with the solar zenithal angle, as well as with the constituents of the atmosphere.

The spectral distribution of solar radiation at ground has a peak around 500 nm. Irradiance is zero at wavelengths less than 280 nm due to absorption by stratospheric ozone (O_3) in these wavelengths. It is attenuated at short wavelengths, from 280 to around 750 nm, due to the scattering by air molecules. The shorter the wavelength, the more intense the scattering. Absorption lines by dioxygen (O_2) and water vapor (H_2O) located between 700 and 800 nm very strongly attenuate the radiation. Other absorption lines, more or less wide, more or less deep, are present beyond 900 nm. They are due to the main gases such as dioxygen (O_2), water vapor (H_2O), carbon dioxide (CO_2), and methane (CH_4). Between 1350 and 1450 nm, and between 1800 and 1950 nm, the incident solar radiation is absorbed by the water vapor. The spectral distributions of the direct and diffuse components of the radiation are very different.

The presence of clouds induces little modification in the shape of the spectral distribution compared to the clear-sky case. As a first approximation, scattering by clouds is non-selective; that is, it affects all wavelengths except the longest in the infrared in a uniform manner. The shape of the spectral distribution is unchanged by the presence of clouds; only the intensity of the attenuation changes.

The range covered by the spectral distribution is not very wide. About 99 % of the solar radiation received at ground is between 250 and 4000 nm. Most of it comes from the contribution of the visible range [380, 780] nm. This proportion is around 55 % for a clear atmosphere and increases with the optical depth of the clouds in cloudy conditions. Likewise, photosynthetically active radiation (PAR), in the range [400, 700] nm, contributes about 45 % in the case of a clear atmosphere. This contribution increases with the optical depth of the clouds. As for the typical pyranometer interval, [330, 2200] nm, the relative contribution of the radiation in this interval to the total radiation is around 98 % and depends little on the sky conditions.

In this chapter, I discuss the spectral distribution of the solar radiation received on a horizontal surface at ground. As with the total radiation, the spectral radiation under clear-sky conditions is an upper bound of that under cloudy conditions at all wavelengths. The clouds attenuate the clear-sky radiation except particular cases of momentary concentration of the solar rays on the faces of the clouds. This is why I start with the case of the cloudless atmosphere before treating the case of the cloudy atmosphere. I present spectral distributions of the irradiance and clearness index as well as their variations with the solar zenithal angle, the turbidity of the clear atmosphere, and the properties of the clouds. I deal with the case of the global irradiance and its direct and diffuse components. For the sake of simplicity in presentation, I limit the discussion in this chapter to the case of the horizontal plane. Reader, I leave it to you to deduce from the knowledge presented the spectral distribution of the irradiance for your own case of an inclined plane.

Given the very wide variety of atmospheric situations and conditions discussed in the previous chapter, I cannot give accurate values of the spectral distribution as I did for the extraterrestrial solar radiation. Consequently, I chose a presentation made mainly of illustrations of typical cases, carried out by numerical simulation with the essential aim of making understand the influence of the atmosphere constituents on the spectral distribution.

For this purpose, I used the libRadtran numerical model that simulates the radiative transfer in the atmosphere. A numerical radiative transfer model is software, or an application, which implements the equations of radiative transfer from atmospheric conditions stipulated by the user. These conditions are described by a series of variables, the inputs, which are very often those listed in the previous chapter as the most influential on the radiation received on the ground. One finds, for example, the date, the solar zenithal angle θ_S, the spectral interval of interest, the elevation of the ground above the mean sea level, the vertical profiles of temperature, density, pressure and volume mixing ratio of the absorbing gases excluding ozone and water vapor, the atmospheric column contents of ozone and water vapor, the types of aerosols and their concentration or their optical depth at several wavelengths, and for the clouds, their optical depth, the phase, the altitudes of the base and the top of the cloud, the liquid or solid water content, and the effective diameter of the droplets and crystals, and finally, the reflection factor of the ground. When these inputs are not completed, the numerical radiative transfer model adopts default values.

In these simulations, the atmosphere is fictitious. Indeed, it is supposed to extend infinitely horizontally, what is called an infinite plane atmosphere. All the atmospheric layers, cloudy or not, are parallel, between two altitudes that do not vary. Each layer is homogeneous vertically and horizontally, in thickness and content. The libRadtran simulator is capable of more complex simulations; it was my own choice to adopt this simple vision, more conducive to the purpose of this book.

For the simulations in this chapter, I arbitrarily chose for fixed inputs, the day 06-19 (19 June), i.e., the 170th day of the year, and a horizontal plane located at the mean sea level. Vertical profiles of temperature, density, pressure, and volume mixing ratio of absorbing gases excluding ozone and water vapor, are typical of midlatitudes in summer. Clouds are made up of water droplets. I will not go into more detail on the inputs because the objective of these simulations is illustration and only illustration.

7.1 Spectral distribution of the irradiance in a cloud-free atmosphere

7.1.1 Influence of the turbidity

Figure 7.1 illustrates the influence of the clear atmosphere on the spectral distribution of solar irradiance received on the ground, at wavelengths between 200 and 2300 nm. Two cases of clear atmosphere are presented. One is limpid, i.e., with a low aerosol load. The other is turbid, i.e., with a large aerosol load. The spectral distribution of extraterrestrial irradiance $E_{0\lambda}$ is also plotted for comparison. The solar zenithal angle is 30°. The spectral distribution gives the irradiance at each wavelength λ. The irradiance integrated over an interval $[\lambda_1, \lambda_2]$ is given by the surface under the curve between these two wavelengths.

The spectral irradiance received at ground (gray curves) exhibits a shape similar to that of the extraterrestrial irradiance $E_{0\lambda}$ with a peak around 500 nm. It is less than that at the top of the atmosphere because of the depletion of the radiation by the atmosphere. The greater the turbidity, the more important the scattering, and the smaller the spectral irradiance on the ground at all wavelengths. The corresponding spectral distributions of the clearness index are drawn in Figure 7.2. If there was no atmospheric attenuation, the clearness index would be equal to 1 at each wavelength. As expected, the clearness index for the clear-sky turbid case is less than or equal to that for the clear-sky limpid case at each wavelength.

The attenuation of solar radiation by the atmosphere is displayed by the difference between the black and gray curves in Figure 7.1, or in an equivalent manner, by the difference between 1 and the corresponding clearness index in gray in Figure 7.2. In the

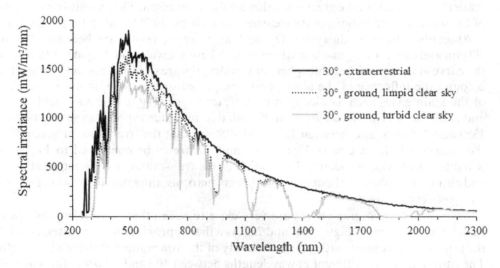

Figure 7.1 Typical spectral distributions of the solar irradiance received on horizontal surfaces located one at the top of the atmosphere and the two others at ground in clear-sky conditions respectively limpid and turbid, between 200 and 2300 nm. Solar zenithal angle is 30°. Results from the numerical code libRadtran simulating the radiative transfer in the atmosphere.

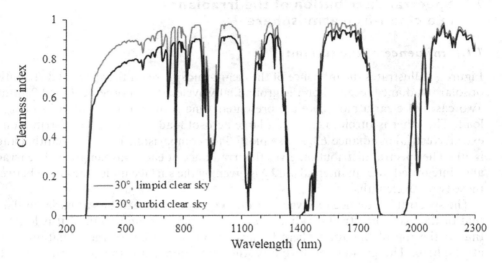

Figure 7.2 Typical spectral distributions of the clearness index for two cloudless atmospheres respectively limpid and turbid between 200 and 2300 nm. Solar zenithal angle is 30°. Results from the numerical code libRadtran simulating the radiative transfer in the atmosphere.

two cases of turbidity presented, this attenuation is not spectrally constant. Solar radiation at wavelengths λ less than 280 nm is absorbed by stratospheric ozone (O_3), and the spectral irradiance received on the ground is zero. The radiation in the short wavelengths is strongly attenuated by the scattering by the air molecules, with, I recall, a scattering intensity all the greater the shorter the wavelength. This results in a rounding of the spectral distributions of the clearness index between 280 and 750 nm (Figure 7.2).

Absorption lines by dioxygen (O_2) and water vapor (H_2O) are between 700 and 800 nm and cause a very marked attenuation. There are wide and deep deviations from the curve at the top of the atmosphere at wavelengths greater than about 900 nm. Here, is found the influence of the absorption lines, more or less wide, more or less deep, of the main gases such as dioxygen (O_2), water vapor (H_2O), carbon dioxide (CO_2), and methane (CH_4). The absorption of radiation by water vapor is so strong between 1350 and 1450 nm, and between 1800 and 1950 nm that the irradiance at ground and the clearness index are zero. Figures 7.1 and 7.2 are to be compared to Figure 4.2 (Chapter 4) showing a spectral distribution of the transmittance for a moderately dry and clean atmosphere without aerosols. Except aerosols, the same absorption and scattering effects are seen.

Note that the two gray curves are very close to each other at wavelengths greater than about 1100 nm, in Figures 7.1 and 7.2. As a first approximation, the attenuation by the atmosphere depends little on the turbidity of the atmosphere at these wavelengths. The situation is very different at wavelengths between 400 and 1100 nm, for which the differences between the two gray curves are large. This is due to the importance of the phenomenon of scattering in this interval. There is a greater influence of turbidity on the spectral irradiance and the spectral clearness index because an increase in turbidity mainly corresponds to an increase in the number of scatterers.

7.1.2 Spectral distribution of the direct and diffuse components

Figure 7.3 exhibits examples of the spectral distributions of the global irradiance received at ground on a horizontal plane and of its direct and diffuse components. The simulation conditions are the same as those in Figure 7.1, i.e., a solar zenithal angle of 30° and two clear atmospheres: a limpid (graph on the left) and a turbid (graph on the right).

When the sky is limpid (graph on the left), the direct component is greater than the diffuse at all wavelengths. The diffuse component is the greatest at wavelengths less than about 750 nm, at which scattering by air molecules is important. The spectral distribution of the diffuse component has a peak between 350 and 500 nm, then decreases rapidly. The spectral distribution of the direct component is quite similar to that of the global irradiance. The two do not match at the shortest wavelengths, where the diffuse component is not negligible. They merge at wavelengths greater than 1100 nm.

The intensity of the scattering phenomenon increases with the turbidity (graph on the right) because the density of the number of scatterers increases. Consequently, the diffuse component increases with turbidity to the detriment of the direct component since the global irradiance decreases with turbidity (see Figure 7.1). The spectral distribution of the diffuse component has a peak between 350 and 500 nm, then decreases rapidly. The direct component is almost extinguished at very short wavelengths. The two components are equal around 700 nm in this simulation. The direct component predominates at wavelengths greater than 700 nm.

Figure 7.4 shows the spectral distribution of the clearness index, for the direct and diffuse components for the two cloudless atmospheres, between 200 and 2300 nm. I recall that the spectral distributions of the clearness indices but for the global irradiance are plotted in Figure 7.2.

Like the global irradiance, the direct spectral clearness index decreases as the turbidity increases, while on the contrary, the diffuse spectral clearness index increases. This corresponds to the fact that the diffuse component is greater in the turbid case than in the limpid case (Figure 7.3). When the turbidity increases further, for example, in the case of a dust storm, the direct component becomes zero. The radiation then has

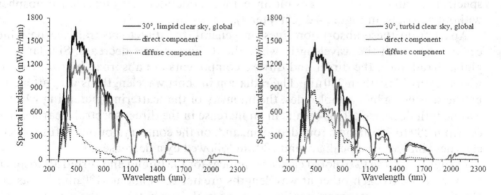

Figure 7.3 Typical spectral distributions of the global irradiance and its components received on a horizontal surface at ground for two cloudless atmospheres: (a) a limpid and (b) a turbid, between 200 and 2300 nm. Solar zenithal angle is 30°. Results from the numerical code libRadtran simulating the radiative transfer in the atmosphere.

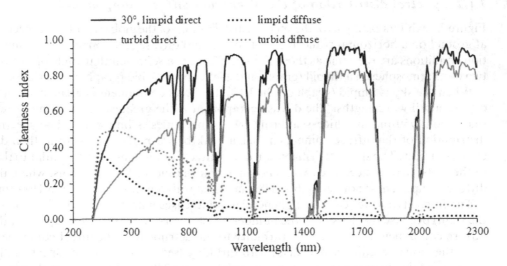

Figure 7.4 Typical spectral distributions of the direct and diffuse clearness indices for two cloudless atmospheres respectively limpid and turbid, between 200 and 2300 nm. Solar zenithal angle is 30°. Results from the numerical code libRadtran simulating the radiative transfer in the atmosphere.

only a diffuse component, which will decrease when the turbidity will further increase. The radiation may become zero. This is what happens in sandstorm episodes, during which darkness settles in gradually and then disappears in the same way.

The spectral distributions of direct and diffuse clearness indices exhibit the same absorption lines as noted above: dioxygen (O_2) and water vapor (H_2O) between 700 and 800 nm, and at wavelengths greater than about 900 nm, absorption lines more or less wide, more or less deep, of the main gases such as dioxygen (O_2), water vapor (H_2O), carbon dioxide (CO_2), and methane (CH_4). As mentioned above, apart from aerosols, the absorption and scattering effects are the same as those present in the spectral distribution of the transmittance, for a moderately dry and clean atmosphere, without aerosols, in Figure 4.2 (Chapter 4).

Apart from these absorption lines, in general, the direct spectral clearness indices increase with the wavelength, while the diffuse indices decrease. Similar to the global irradiance, the direct and diffuse components are absorbed by ozone (O_3) at wavelengths λ less than 280 nm. The radiation in short wavelengths is very attenuated by the scattering by air molecules, the intensity of the scattering increasing when the wavelength decreases. This results in an increase in the direct spectral clearness indices from 280 to 750 nm in a rounded form, and, on the contrary, for the diffuse spectral clearness indices by a peak around 280 nm followed by a decrease.

In Figure 7.2, the two global clearness indices for the limpid and turbid atmospheres are very close to each other at wavelengths greater than about 1100 nm: It was concluded that as a first approximation, the spectral distribution of the clearness index depends little on the turbidity of the atmosphere at these wavelengths. Figure 7.4 details this observation. It can be seen that the difference between the two direct spectral clearness indices is compensated by the difference, in opposite directions, between the two diffuse clearness indices.

7.1.3 Influence of the solar zenithal angle

The previous figures show how much the attenuation depends on the wavelength λ. The attenuation and its spectral distribution also strongly depend on the solar zenithal angle θ_S. The larger θ_S, the larger the air mass, the more the influences of absorption and scattering are felt, and the weaker the radiation received on the ground. Figure 7.5 illustrates the spectral distribution of the irradiance E_{clear} received on the ground for three solar zenithal angles: 0°, 30°, and 60°, in the case of a cloudless and limpid atmosphere. The three spectral distributions in Figure 7.5 have the same shape. The spectral irradiance is all the smaller as θ_S is large. The same absorption lines and the same scattering effects are found for each angle. The intensity of the spectral attenuation is not simply proportional to $\cos(\theta_S)$. If this were the case, the ratio between the curves, for example, 0° and 60°, should be the same as the ratio of cosines and should be constant with λ. For example, since $\cos(60°) = 1/2 = \cos(0°)/2$, the curve for 60° should be half that for 0°, at any λ. However, it is not. The curve for 60° is less than half that for 0°. This shows that the influence of θ_S on attenuation is not simply proportional to $\cos(\theta_S)$ and is also a function of λ.

To further illustrate this and other points, Figure 7.6 shows the spectral distribution of the clearness index K_T corresponding to the spectral irradiances in Figure 7.5 for each of the three solar zenithal angles. If the atmosphere were completely transparent, K_T would be equal to 1 at each wavelength. The larger θ_S, the greater the air mass, the more the influences of absorption and scattering are felt, and the smaller the clearness index K_T. The clearness index is small at the shortest wavelengths due to the absorption by ozone as seen in Figure 7.1. K_T increases with λ as the intensity of scattering by the molecules of the air decreases. Of course, the same absorption lines and the same scattering effects as a function of λ are the same as in

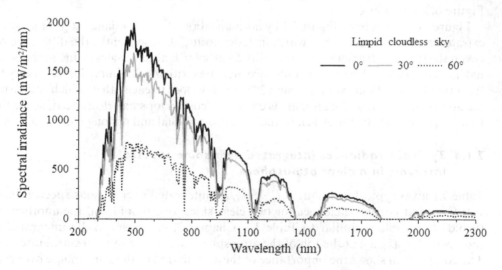

Figure 7.5 Typical spectral distributions of the solar irradiance received on a horizontal surface at ground in cloudless limpid conditions between 200 and 2300 nm for three solar zenithal angles: 0°, 30°, and 60°. Results from the numerical code libRadtran simulating the radiative transfer in the atmosphere.

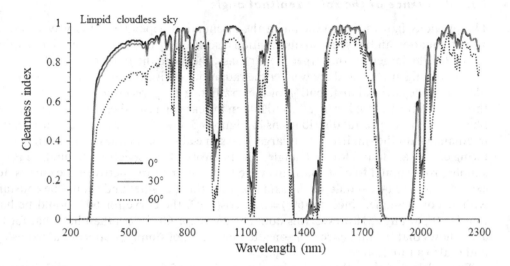

Figure 7.6 Typical spectral distributions of the clearness index K_T between 200 and 2300 nm for three solar zenithal angles: 0°, 30°, and 60°, for a cloudless limpid atmosphere. Results from the numerical code libRadtran simulating the radiative transfer in the atmosphere.

Figure 7.1. In limpid conditions, as in this simulation, K_T can reach very large values close to 1 at wavelengths where the atmosphere can be considered almost transparent, at least when θ_S is small. Indeed, K_T decreases when θ_S increases, as also seen in Figure 6.3 (Chapter 6).

Figure 7.6 derives from Figure 7.1 by normalization of the irradiance at ground by the extraterrestrial irradiance $E_{0\lambda}$, which includes $\cos(\theta_S)$. Consequently, the difference between the spectral distributions for the different solar zenithal angles is due to the residual influence of θ_S. This residual influence increases with θ_S: The curve for 60° is very far from that for 0°. The curves for 0° and 30° are very close to each other, which can justify the approximation that the air mass is equal to $1/\cos(\theta_S)$ for small θ_S. Indeed, according to this approximation, the clearness indices must be equal and no longer depend on θ_S.

7.1.4 *Typical irradiances integrated over some spectral intervals in a clear atmosphere*

Table 7.1 gives typical irradiances $E_{clear}(\lambda_1, \lambda_2)$ integrated over various spectral intervals $[\lambda_1, \lambda_2]$ at normal incidence for two clear skies, one limpid and the another one turbid. This table is similar to Table 3.3 (Chapter 3) giving spectrally integrated irradiances $E_{TSI}(\lambda_1, \lambda_2)$ at the top of the atmosphere over these same spectral intervals. The comparison shows the importance of the attenuation of the atmosphere over each spectral interval, as seen with Figures 7.1 and 7.2.

As already seen, $E_{clear}(\lambda_1, \lambda_2)$ in a given interval decreases as the turbidity increases. As noted above, this decrease is not constant over wavelengths. This can be illustrated by calculating the ratio between the irradiances for the two turbidities in Table 7.1. For example, the ratio between the total irradiances for limpid and turbid conditions

Table 7.1 Typical irradiances (W m^{-2}) received at ground at normal incidence integrated over various spectral intervals for two cloudless atmospheres: one limpid and one turbid

Spectral interval (nm)	Limpid	Turbid
250–20,000	1051	972
250–4000	1045	966
380–2100	969	898
400–1100	798	735
1100–4000	184	178
1100–20,000	190	184
4000–20,000	6	6
250–400	71	61
400–800	595	544
280–315 (UV-B)	3	3
315–400 (UV-A)	70	61
480–485 (blue)	18	17
510–540 (green)	59	54
620–700 (red)	122	113
380–780 (visible CIE)	591	539
400–700 (PAR)	484	441
330–2200 (typical spectral range of pyranometers)	1017	940

PAR stands for photosynthetically active radiation. CIE stands for International Commission on Illumination. Results from the numerical code libRadtran simulating the radiative transfer in the atmosphere.

(Table 7.1, first line) is approximately 0.92 (= 972/1051). It is 0.97 in the [1100, 20,000] nm range (= 184/190) and 0.91 in the visible range (= 539/591).

The UV-B irradiance, in the interval [280, 315] nm, is very low: a few W m^{-2} (Table 7.1). It is the same under the two clear sky conditions in this simulation as it mainly depends on the concentration of atmospheric ozone. The irradiance in UV-A ([315, 400] nm) is significantly greater by about 20 times and is sensitive to the turbidity of the atmosphere.

In addition, Table 7.2 quantifies the relative contribution of spectrally integrated $E_{clear}(\lambda_1, \lambda_2)$ to the total irradiance at normal incidence for the two clear-sky conditions. For the sake of comparison, I also reported the fractions compared to the extraterrestrial total irradiance already presented in Table 3.3 (Chapter 3). About 99 % of the total irradiance E_{clear} received on ground is in the range [250, 4000] nm; the share of the interval [4000, 20,000] nm is very small (Table 7.2). In general, the longest wavelengths, greater than 900 nm, have the broadest absorption lines. Consequently, the relative contributions of the longer wavelengths are lower at ground level than at the top of the atmosphere. Correlatively, the relative contributions of the shorter wavelengths are greater at ground level than at the top of the atmosphere. For example, the irradiance $E_{clear}(\lambda_1, \lambda_2)$ in the interval [400, 1100] nm accounts for approximately 76 % of E_{clear}, while at the top of the atmosphere the contribution of this interval is only 67 % of E_{TSI}.

The contribution of UV-B radiation to the total irradiance E_{clear} is very small, less than 1 % of E_{clear}. That of UV-A radiation is much larger, around 7 % of E_{clear}. These proportions for ultraviolet at ground level are close to those observed at the top of atmosphere. More than half (55 –56 %) of E_{clear} comes from the contribution of the interval [380, 780] nm, which is the visible range defined by the International Commission on Illumination. The contribution of the visible range is much lower at the top of the

Table 7.2 Typical values of the fraction (%) of the irradiance received at normal incidence at the top of the atmosphere (left column) and on the ground (right columns) in various spectral ranges relative to the respective total irradiance in two cloudless conditions

Spectral interval (nm)	Top of the atmosphere	Limpid	Turbid
250–20,000	100	100	100
250–4000	99	99	99
380–2100	89	92	92
400–1100	67	76	76
1100–4000	25	18	18
1100–20,000	25	18	19
4000–20,000	1	1	1
250–400	8	7	6
400–800	49	57	56
280–315 (UV-B)	1	<1	<1
315–400 (UV-A)	6	7	6
480–485 (blue)	2	2	2
510–540 (green)	5	6	6
620–700 (red)	10	12	12
380–780 (visible CIE)	49	56	55
400–700 (PAR)	40	46	45
330–2200 (typical spectral range of pyranometers)	93	97	97

PAR stands for photosynthetically active radiation. CIE stands for International Commission on Illumination. Fractions are rounded up to the nearest integer. Results from the numerical code libRadtran simulating the radiative transfer in the atmosphere.

atmosphere (49 %). Photosynthetically active radiation (PAR), in the range [400, 700] nm, contributes in this example to around 45 %–46 % of E_{clear}, more than at the top of the atmosphere (40 % of E_{TSI}). As for pyranometers, if they do not measure all of the total irradiance E_{clear}, they capture most of it, around 97 %.

7.2 Spectral distribution of the irradiance in a cloudy atmosphere

Clouds are made up of many water droplets or ice crystals, or a combination of both. The intensity of the scattering is very strong, and the radiation directed toward the ground can be strongly attenuated when passing through a cloud. The greater the optical depth of the cloud, the greater the attenuation. There is a great diversity of clouds and therefore a great diversity of effects. However, the general properties of these effects can be presented with the help of simple simulations.

7.2.1 Influence of the cloud optical depth

Figure 7.7 illustrates the influence of the optical depth of the cloud on the spectral distribution of the solar radiation, at wavelengths between 200 and 2300 nm. The receiving surface on the ground is horizontal, and the solar zenithal angle is 30°. Three cases are simulated: a cloud layer of optical depth 5, another of optical depth 15, and, for comparison, a limpid cloudless atmosphere. The corresponding clearness indices are plotted in Figure 7.8.

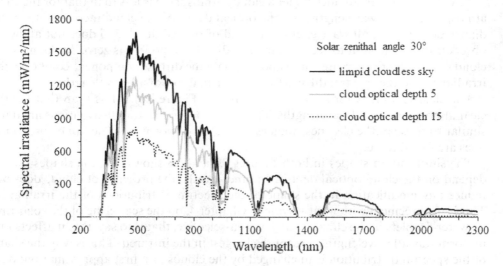

Figure 7.7 Typical spectral distributions of the solar irradiance received on a horizontal surface located at ground in limpid clear-sky conditions, and in cloudy conditions with cloud optical depths of 5 and 15, between 200 and 2300 nm. Solar zenithal angle is 30°. Results from the numerical code libRadtran simulating the radiative transfer in the atmosphere.

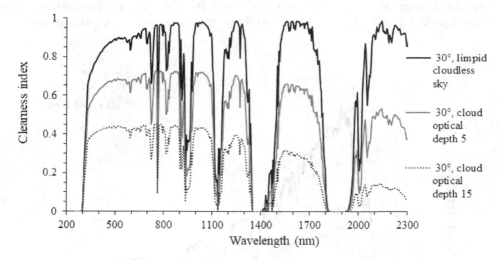

Figure 7.8 Typical spectral distributions of the clearness index in limpid clear-sky conditions, and in cloudy conditions with cloud optical depths of 5 and 15, between 200 and 2300 nm. Solar zenithal angle is 30°. Results from the numerical code libRadtran simulating the radiative transfer in the atmosphere.

The three curves in Figure 7.7 have the same shape: zero irradiance at wavelengths λ less than 280 nm, strong growth when λ increases to around 500 nm, and then a slow decay, marked by different absorption lines. The different absorption lines are the same as for the clear atmosphere. They have been detailed previously, and I do not repeat them.

As expected, the irradiance under a cloudy atmosphere is less than that for the clear atmosphere, at all wavelengths. As the optical depth of the cloud increases, the irradiance decreases at all wavelengths. A cloud of optical depth 3–5 does not allow an observer on the ground to see the sun: The direct component is zero. As soon as the cloud optical depth is greater than about 3, only the diffuse component composes the irradiance and the spectral diffuse fraction is equal to 1 at all wavelengths.

The clearness index for a cloudy atmosphere (Figure 7.8) is less than that for the clear atmosphere, at all wavelengths. The shapes of the spectral distributions are quite similar to those of the clearness indices for a clear atmosphere. The same absorption lines are found there.

The similarity in shapes in both Figures 7.7 and 7.8 shows that the shapes do not depend on the cloud optical depths. In other words, the presence of clouds does not induce any modification in the shape of the spectral distributions of the irradiance. This is in agreement with what I wrote in Chapter 4, on the scattering of the solar rays by water droplets in the clouds, which is non-selective, that is to say, that it affects in a uniform way all wavelengths, except the longest in the infrared. This is why the shape of the spectral distribution is unchanged by the clouds as a first approximation; only the intensity of the attenuation changes.

7.2.2 Influence of the solar zenithal angle

Figure 7.9 shows the spectral distribution of the irradiance E_{cloud} received at ground for three solar zenithal angles θ_S: 0°, 30°, and 60°, for a cloudy atmosphere, with a cloud optical depth set to 5. Figure 7.10 shows the corresponding clearness indices.

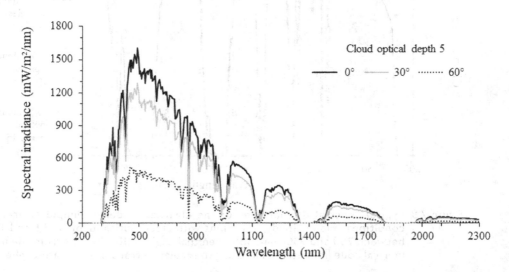

Figure 7.9 Typical spectral distributions of the solar irradiance received on a horizontal surface located at ground in cloudy conditions, between 200 and 2300 nm, for three solar zenithal angles: 0°, 30°, and 60°. The cloud optical depth is 5. Results from the numerical code libRadtran simulating the radiative transfer in the atmosphere.

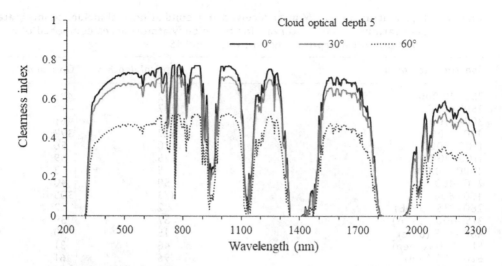

Figure 7.10 Typical spectral distributions of the clearness index K_T in cloudy conditions, between 200 and 2300 nm, for three solar zenithal angles: 0°, 30°, and 60°. The cloud optical depth is 5. Results from the numerical code libRadtran simulating the radiative transfer in the atmosphere.

The spectral distributions in Figure 7.9, respectively Figure 7.10, have the same general shape. The spectral irradiance, respectively the spectral clearness index, decreases when θ_S increases at each wavelength. These graphs illustrate in particular the great importance of the solar zenithal angle on the attenuation of radiation, including in a cloudy atmosphere.

7.2.3 Typical irradiances integrated over spectral intervals in cloudy conditions

Table 7.3 gives some typical irradiances $E_{cloud}(\lambda_1, \lambda_2)$ integrated over various spectral intervals $[\lambda_1, \lambda_2]$ at normal incidence for two atmospheres composed of a cloud of optical depth of, respectively 5 and 15. This table is similar to Table 7.1 giving spectrally integrated irradiances but for clear-sky conditions. In addition, Table 7.4 quantifies the relative contribution of spectrally integrated $E_{cloud}(\lambda_1, \lambda_2)$ to the total irradiance E_{cloud} at normal incidence for the two cases of cloudy sky. For the sake of comparison, I also reported the fractions compared to the total radiation E_{TSI} at the top of the atmosphere, already presented in Table 3.3 (Chapter 3).

The irradiance $E_{cloud}(\lambda_1, \lambda_2)$ in a given interval decreases as the optical depth of the cloud increases (Table 7.3). This decrease is not constant over wavelengths. As done previously for Table 7.1, this can be illustrated by calculating the ratio between the irradiances for the two optical depths in Table 7.3. For example, the ratio between the total irradiances for the two optical depths (Table 7.3, first line) is about 0.61 (= 493/802). The ratio is also 0.61 (= 490/797) for the interval [250, 4000] nm (Table 7.3, second line). It is around 0.62 (= 484/784) for the typical spectral band of pyranometers (Table 7.3, last line). As a whole, this ratio remains approximately constant at wavelengths shorter

Table 7.3 Typical irradiances (W m^{-2}) received at ground at normal incidence integrated over various spectral intervals for two cloudy atmospheres composed of a cloud of optical depth of respectively 5 and 15

Spectral interval (nm)	Optical depth 5	Optical depth 15
250–20,000	802	493
250–4000	797	490
380–2100	747	461
400–1100	627	398
1100–4000	120	56
1100–20,000	125	59
4000–20,000	5	3
250–400	58	40
400–800	473	305
280–315 (UV-B)	2	2
315–400 (UV-A)	57	39
480–485 (blue)	15	10
510–540 (green)	48	31
620–700 (red)	96	61
380–780 (visible CIE)	470	304
400–700 (PAR)	387	251
330–2200 (typical spectral range of pyranometers)	784	484

PAR stands for photosynthetically active radiation. CIE stands for International Commission on Illumination. Results from the numerical code libRadtran simulating the radiative transfer in the atmosphere.

than around 2000 nm. It can be concluded that as a first approximation, the effects of clouds are spectrally neutral at wavelengths shorter than around 2000 nm. In other words, the depletion by the clouds does not depend on the wavelength at wavelengths shorter than around 2000 nm as a first approximation. At greater wavelengths, the ratio is less than 0.6 and depends on the interval.

As for the clear sky, approximately 99 % of the total irradiance E_{cloud} received at ground is in the interval [250, 4000] nm (Table 7.4); the share of the interval [4000, 20,000] nm is very small. In general, since wavelengths greater than 900 nm have the broadest absorption lines, the relative contributions of the longest wavelengths are less at ground level than at the top of the atmosphere. Correlatively, the relative contributions of wavelengths less than 900 nm are greater at ground level than at the top of the atmosphere. For example, the irradiance $E_{cloud}(\lambda_1, \lambda_2)$ in the interval [380, 2100] nm accounts for approximately 94 % of E_{cloud}, while at the top of the atmosphere the contribution of this interval is only 89 % of E_{TSI}. Likewise, the irradiance $E_{cloud}(\lambda_1, \lambda_2)$ in the interval [400, 1100] nm accounts for approximately 80 % of E_{cloud}, while the fraction is only 67 % of E_{TSI} at the top of the atmosphere.

The irradiance in the UV-B, in the range [280, 315] nm, is very low, around 2 W m^{-2} (Table 7.3). There is very little variation in the attenuation in UV-B depending on the optical depth of the clouds. According to Table 7.1, the UV-B irradiance is 3 W m^{-2} for clear and turbid cloudless atmospheres. The influence of the constituents of the tropospheric atmosphere on UV-B radiation is therefore weak; it is the content of the stratospheric ozone layer, between 20 and 40 km above sea level that matters most. The UV-B irradiance represents less than 1 % of the total irradiance E_{cloud} (Table 7.4). The UV-A irradiance ([315, 400] nm) is much greater. It is a few tens of W m^{-2} (Table 7.3) and

Table 7.4 Typical values of the fraction (%) of the irradiance received at normal incidence at the top of the atmosphere (left column) and on the ground (right columns) in various spectral ranges relative to the respective total irradiance under cloudy conditions

Spectral range (nm)	Top of the atmosphere	Optical depth 5	Optical depth 15
250–20,000	100	100	100
250–4000	99	99	99
380–2100	89	93	94
400–1100	67	78	81
1100–4000	25	15	11
1100–20,000	25	16	12
4000–20,000	1	1	1
250–400	8	7	8
400–800	49	59	62
280–315 (UV-B)	1	<1	<1
315–400 (UV-A)	6	7	8
480–485 (blue)	2	2	2
510–540 (green)	5	6	6
620–700 (red)	10	12	12
380–780 (visible CIE)	49	59	62
400–700 (PAR)	40	48	51
330–2200 (typical spectral range of pyranometers)	93	98	98

PAR stands for photosynthetically active radiation. CIE stands for International Commission on Illumination. Fractions are rounded up to the nearest integer. Results from the numerical code libRadtran simulating the radiative transfer in the atmosphere.

represents around 7 % of E_{cloud} (Table 7.4). These proportions for ultraviolet are close to those observed at the top of the atmosphere. The radiation in UV-A is weaker in the presence of cloud than in cloudless conditions and decreases as the optical depth of the cloud increases.

If the irradiance in the visible range [380, 780] nm decreases when the cloud optical depth increases, on the other hand, the relative contribution of this interval to E_{cloud} increases with the optical depth. It is 59 % and 62 %, respectively, in the simulations here, much more than that at the top of the atmosphere (49 %). In general, the relative contributions of the intervals to E_{cloud} increase as the cloud optical depth increases. For example, the PAR in the interval [400, 700] nm contributes in these simulations to 48 % and 51 %, respectively of E_{cloud}. This is more than for the clear sky (46 % and 45 % of E_{clear}, Table 7.2) and much more than at the top of the atmosphere (39 % of E_{TSI}).

For very large intervals, the contributions relating to E_{cloud} are similar to those relating to E_{clear}. For example, for the interval [380, 2100] nm, the contributions relating to E_{cloud} are 93 % and 94 % and similar to those relating to E_{clear} (92 %). Another example is that of the typical spectral range of pyranometers, i.e. [330, 2200] nm. The contributions relating to E_{cloud} (98 %) and E_{clear} (97 %) are very close: These instruments capture a very large part of the total radiation and do so homogeneously for all types of sky.

Chapter 8

Variability – implications for estimating radiation

At a glance

Solar radiation fluctuates over time and space. The temporal and spatial scales, also called characteristic scales, indicate the orders of magnitude of the sizes of these fluctuations. Meteorological variables are often considered to be random variables. It is possible to measure different values under the same meteorological conditions even if the noise of the measurement is negligible, because there are fluctuations at smaller scales which cause the mean value to fluctuate on the scale of the observation, whatever that scale. A time series of measurements describes the mean values and the fluctuations over time; an image is a spatial distribution of measurements and describes the mean values and the fluctuations in space. According to the ergodicity hypothesis, there is a kind of equivalence between a time series measured at a geographic site and a set of pixels acquired by a spaceborne imager, and the time averages are the same as the spatial averages.

An instrument typically consists of a sensor or detector and a digital acquisition system with computation and storage capabilities. The instrument samples the meteorological variable; that is, it takes at often user-defined, usually regular intervals the values of the signal observed over a certain period of time or surface. The measurement is the integration of the variable over this period, called duration or integration period. In the case of an image, the integration is spatial and carried out on a pixel. This notion of pixel can be generalized to the gridded outputs of numerical meteorological or climatic models or other models.

The sampling is regular if the measurements are all acquired over the same time duration or pixel size and follow each other without interruption. Otherwise, it is irregular. The time interval or the distance separating the beginnings or ends of two consecutive measurements is called the sampling step. The smaller the sampling step, the smaller the size of the finest observable variations. According to the sampling theorem, the sampling must be regular, with a step less than half of the smallest size of interest for a good discrete representation of the properties of solar radiation estimated from the measurements. Conversely, this theorem defines the size of the smallest detectable variations in a series of measurements given by third parties.

Solar radiation, like all other meteorological variables and many others, exhibits fluctuations in time and space, of different durations and sizes. These fluctuations around average values constitute the variability of the radiation. I have shown a few examples of variability in previous chapters, both for radiation at ground and for radiation received at the top of the atmosphere.

The purpose of this chapter is to give some information on this variability, then to describe some of the implications for measuring radiation and meteorological variables in general. The purpose of the measurements is to quantify the radiation, its average values, and its fluctuations in time and space. Consequently, the measurement acquisition processes, the digital processing of the measurements, and all the methods for estimating radiation must take into account the variability of the radiation.

I introduce first the time and space scales, the random nature of the radiation received at ground, and ergodicity that allows the comparison between temporal and spatial statistics. These are very useful concepts to describe the variations of this radiation. I then define the measurement, the sampling, and the information support both in time (the integration period) and in space (the size of a cell, or the pixel). Finally, I present the sampling theorem that defines optimal sampling for a good discrete representation of the properties of solar radiation estimated from the measurements, as well as the size of the smallest details detectable in series of measurements acquired by third parties.

8.1 Time and space scales

It is practical in meteorology and other sciences to describe phenomena and therefore the variables describing these phenomena, using time and space scales. They give orders of magnitude of the sizes characteristic of the fluctuations of the variables in time and space. For example, a thunderstorm has a finite spatial dimension of the order of 10 km and a temporal dimension of the order of a few hours: The scales of variation are therefore 1 h and 10 km. The solar radiation received on the ground exhibits a wide variety of temporal and spatial scales. They range from the second and a millimeter, under the influence of atmospheric turbulence, to the millennium (and beyond) and tens of thousands of kilometers (the perimeter of the Earth being around 40,000 km).

Let me illustrate this notion. Figure 8.1 is a schematic representation of a particular geographic location of two plains separated by a mountain range oriented north–south, 100 km wide. Thus, the relief exhibits a space scale of 100 km. A moisture-laden wind blows from west to east. Clouds form on the windward slope, i.e., the western slope. Clouds themselves exhibit a space scale of 100 km. As the solar radiation received on the ground strongly depends on the clouds, it also has a spatial scale of 100 km in this western part. However, in this diagram, no cloud being present in the eastern part, there is no variation due to clouds and the spatial fluctuations of the radiation are those of the constituents of the clear atmosphere.

Reader, I readily acknowledge my shortcomings in drawing; the quality of this diagram is pretty appalling. However, the situation shown here is not unrealistic. It evokes the Cascade Range or Cascade Mountains, which borders the Pacific coast in the states of British Columbia in Canada, and Washington and Oregon in the United States, or

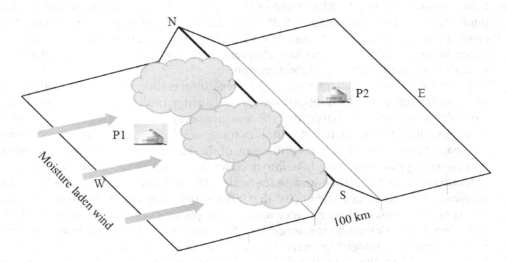

Figure 8.1 Schematic representation of a geographic location with two plains separated by a mountain range oriented north–south, 100 km wide. A moisture-laden wind blows from west to east. Symbols *P1* and *P2* are pyranometers measuring solar irradiation.

even that of the Southern Alps, which borders the west coast of the South Island of the New Zealand.

In general, a scale is an indication of order of magnitude, unless a specific phenomenon is described. For example, a scale of 100 km means that the dimensions of the phenomenon concerned are around 100 km. They can vary from, say, 60 to 500 km. Because of the nature of meteorological phenomena, time and space scales are linked. For example, a timescale of 1 h corresponds to a spatial scale of 100 km. Not all space scales are available for the same timescale and vice versa. For example, there are no phenomena having as characteristic scales the couples (1 min, 1000 km), or (1 day, 1 m).

Three classes of scales in meteorology, or oceanography or climate are often distinguished for practical reasons:

- The global or synoptic scale relates to the general circulation of air or water masses over large areas of the order of 1000 km and timescales of the order of the day or more,
- the mesoscale includes the space scales from 1 to 100 km and timescales from 1 min to 10 h,
- the microscale includes the smaller scales, less than 1 km and 10 min.

National and international meteorological observation networks perform so-called synoptic measurements. If a network is dense enough, i.e., the stations are not distant from their nearest neighbor by more than 50 km, the measurements are called mesoscale.

You can also find another classification in the scientific literature:

- global or synoptic scale whose space scale is of the order of 10,000 km and timescale is the week,
- continental scale whose space scale is of the order of 1000 km and timescale is 1 day,
- regional scale whose space scale is of the order of 100 km and timescale is 1 h,
- urban or local scale whose space scale is of the order of 1 km and timescale is 1 min.

Another classification is proposed by the World Meteorological Organization for horizontal meteorological scales,[1] with a factor two uncertainty:

- planetary scale whose space dimension is greater than 3000 km,
- large scale between 100 and 3000 km,
- mesoscale between 3 and 100 km,
- toposcale or local scale between 100 m and 3 km,
- microscale less than 100 m.

Scales of variations may also be called characteristic scales. For timescales, the term characteristic period or simply period, often denoted T, is sometimes used if the phenomenon shows a periodic trend. For example, the extraterrestrial irradiance at normal incidence E_{0N} exhibits seasonal and annual periods. The frequency is the inverse of the period and is often noted f. For example, if the period T is equal to 1 s, 1 h, and 1 year, respectively, the frequency is equal to $1\,s^{-1}$, $1\,h^{-1}$, and $1\,year^{-1}$. In the space domain, the equivalent of the period is the wavelength. The inverse of the wavelength is the wave number, also called spatial frequency.

Whether in space or in time, given a context of scales spanning several decades, that is to say, several orders of magnitude, the term high frequencies may be used to designate the phenomena, or the part of the phenomena varying quickly (small periods or wavelengths). In a similar way, the term low frequencies may designate the phenomena, or part of the phenomena varying slowly (long periods or wavelengths). These notions of low and high frequencies are therefore relative. For example, given irradiations ranging from hourly to decade (10 year) the intraday and day-to-day variations in irradiation are high frequencies, while the interannual variations are low frequencies. If I am dealing with monthly irradiations, the variations on scales of months to a few years are high frequencies, and those on decadal scales and beyond are low frequencies.

What is this classification used for? Why define characteristic scales? What is their practical significance? The characteristic scales allow comparisons between different phenomena, in particular to include them or not in the current study. By way of illustration, suppose that I am interested in the solar radiation received at ground in

---

1 Guide to Instruments and Methods of Observation. WMO-No. 8, 2018 edition, 2018, World Meteorological Organization, Geneva, Switzerland. [Vol 1 Measurement of Meteorological Variables, Chapter 1 – General.]

monthly average, over a very large geographic area with a dimension greater than a few thousand km. I have to take into account the temporal and spatial variations of the extraterrestrial solar irradiation H_0 since H_0 varies quite strongly with the month and the latitude (Chapter 4). In contrast, suppose that I am interested in the hourly irradiation in a restricted area of 100×100 m². In this case, I can neglect the spatial variations of H_0 because they are very small. I can assume that H_0 is the same everywhere in the area at any time. Thanks to this physical consideration, I can perform only one calculation of H_0 for the whole area at each time step, without loss of precision, which is a benefit from a practical point of view. Another practical benefit is that using these scales, I can better share the notions of fluctuations and variability with you, reader.

8.2 Random nature of meteorological variables – ergodicity

Meteorological variables are often considered as random variables or stochastic variables. That is, given a variable, different values may be measured for the same atmospheric conditions, even if the noise of the measurement is negligible. This happens because, regardless of the observation scale, there are fluctuations at smaller scales which cause the value on the observation scale to fluctuate.

Here is an example to make this random nature more tangible. Suppose I have several identical instruments, without noticeable measurement noise, which I place in a very dense network. So placed next to each other, they measure the same phenomenon at the same times, say every second. I would then observe that the time series of measurements of the different instruments differ from one instrument to another; they are not the same. Microturbulence or the heterogeneity of scattering bodies and their density in the clear or cloudy atmosphere induce fluctuations at high frequencies. As a result, the instantaneous values differ from one moment to another and also from one measurement site to another. When averaging the time series over several hours or more, the high-frequency fluctuations would average identically over space and time and the same averages would be obtained at all sites at these low frequencies.

A series of measurements is a realization of the random process, also called the stochastic process. In theory, the duration of the random process is infinite. However, for obvious practical reasons, a series of measurements has a limited extent in time or space. It is not possible to represent reality perfectly. This difficulty can be accommodated by limiting the analysis of measurements to certain scales, as it will be seen later.

Who says random says that it is necessary to have several realizations to describe the process, even if it is restricted to certain scales. Usually, this is not possible in meteorology for practical reasons. This can happen but on an ad hoc basis, as in the example above of a very dense network of identical instruments, where each time series is a realization of the random process. In the general case, there is only one realization of finite extent to describe the process, which poses a problem.

This is where ergodicity comes in. This property of stochastic processes is quite difficult to describe precisely without resorting to advanced concepts in probability, which I do not want to do here. Let me say that a stochastic process is ergodic if its description can be approximated by the analysis of a single realization, provided it is long enough. If the studied variable, for example, the irradiation, results from an ergodic process, then a single series of measurements is sufficient.

From a practical point of view, in general, the ergodic hypothesis is made when processing a series of measurements. That is to say, it is assumed that the realization available is sufficient to describe the variable studied. You understand well, reader, that in practice, given the finite nature of the series of measurements in time or space, the hypothesis of ergodicity concerns only a set of scales characteristic of a phenomenon and not an infinity of scales. The duration of this series, or its spatial extent, must be long enough in relation to the scales that one wishes to study. For example, if I want to study fluctuations in radiation over the course of a year, it is desirable to have a series of measurements covering several years, at least ten. Likewise, it is desirable to have an image or a radiation map of extent of 1000 km or more to analyze the fluctuations of the mesoscale radiation (100 km).

Users of measurements are not always interested in all the statistical moments to describe the variable: Users are often satisfied with the mean, the variance, or even the covariance. In this case, the ergodic hypothesis is weaker than for a strict ergodicity concerning all the statistical moments.

The ergodic hypothesis is difficult to verify. If a very long series of measurements is available, it is possible to split it into several sub-series and, by calculating several statistics, validate or invalidate this hypothesis depending on the scales studied. Likewise, it is possible to split a spatial distribution of measurements (image) into several sub-images and analyze the variations in statistics between the sub-images and the image.

The ergodic hypothesis should be considered with caution. Consider two simple examples. First, consider a random variable over time, which is a sinusoid with a period of 10 days on which is superimposed high-frequency, low-amplitude white noise. Since the average of white noise is zero, the average of this sinusoid over 10 days is zero. If the series of measurements is longer than this period, lasting 100 days, the mean is always zero and the ergodic hypothesis is verified for the mean. In other words, this series of limited duration well represents the random variable with regard to the mean. For the second example, suppose a second random variable, made up of the first one on which is superimposed a noticeable drift, that is to say, a continuous increase in values over time. The 10-day average thus increases over time. This new 100-day series will not have a zero mean, unlike the random variable, and the ergodic hypothesis is not verified with regard to the mean. These two examples are undoubtedly caricatures, but their purpose is to illustrate certain limitations of the ergodic hypothesis.

It is conceivable that a process comprising fluctuations of large amplitude at very large scales, or a continuous increase, or decrease, in values over time or space, may not be ergodic. It appears that the ergodic hypothesis is better verified when the influence of large-scale fluctuations on the measured signal is weak.

It may be useful to make changes in variables to better respect the ergodic hypothesis. For example, suppose a time series of measurements of the daily irradiation H_{day} received at ground. H_{day} depends on the properties of clouds, the constituents of the cloudless atmosphere, and the extraterrestrial irradiation H_{0day}. Therefore, fluctuations in H_{0day} will modulate the measurements of H_{day}. However, you have seen that it is possible to describe H_{0day} and its fluctuations quite precisely. Thus, H_{0day} is not a random variable and exhibits deterministic fluctuations that may interfere with the ergodic hypothesis. The influence of H_{0day} and its fluctuations can be removed, for example, by using the daily clearness index as a variable instead of the irradiation H_{day}.

Assuming ergodicity has several important practical implications for solar radiation, as well as for other variables.

The first implication is that it only takes one realization to describe the random variable. A time series of measurements makes it possible to describe the average values and fluctuations over time; an image, i.e., a spatial distribution of measurements, makes it possible to describe the mean values and fluctuations in space.

The second implication is important for the observation of the Earth by imagers on board satellites. There is a kind of equivalence between space and time. That is, the observation by a time series is equivalent to that of an image. In other words, the variable observed using a time series can also be described by a set of pixels acquired by a spaceborne imager. In particular, the time averages of the variable are the same as the spatial averages. Let me go back to Figure 8.1 as an example. In this figure, I have schematized two pyranometers, *P1* and *P2*, located one to the west of the mountain range and the other to the east. Suppose the western and eastern parts are each covered by a pixel of 200×200 km². This pixel is the spatial average of the radiation. If the process is ergodic, then the time average of the series of measurements acquired over a few hours at *P1* will be equal to the value of the pixel west of the mountain. Similarly, the time average of the series of measurements acquired over a few hours at *P2* will be equal to the value of the pixel east of the mountain.

Has the ergodicity hypothesis been extensively verified for imagers? No, but it can be considered as de facto verified. Images acquired from satellites have been used in meteorology and oceanography for decades, without any indication that the hypothesis should be called into question.

The equivalence between space and time is not obvious. Indeed, how many pixels are needed for a time series of a given duration for the averages to be equivalent? More generally, given certain timescales, what are the equivalent spatial scales? A distance is equal to a duration multiplied by a speed. So what is the speed allowing this equivalence? The answer is not obvious as illustrated by an example, shown in Figure 8.2. Suppose an infinite set of small clouds all identical, of small vertical size, rather circular and of diameter 100 m, separated by approximately 500 m, and moving at constant altitude. Suppose these clouds are distributed randomly over a surface of infinite length and constant width equal to, say, 30 km. Suppose the set of clouds is rigid; that is, the clouds always stay in the same position relative to each other. This is sometimes called the frozen turbulence hypothesis: The relative positions do not change. This situation may seem very schematic. However, it is not unrealistic as it may evoke the cloud fields observed over the great plains in Europe in summer, which are made up of many small cumulus clouds, called fair weather cumulus clouds.

Suppose an imager on board a satellite, which has pixels of 10×10 km². Assume a pyranometer at ground, located under the moving axis of this set of clouds (Figure 8.2). If the whole has a uniform moving speed of 10 km h⁻¹, always in the same direction, then the pyranometer sees about 10 km of this set of clouds in 1 h, i.e., the size of the pixel. The time average over 1 h corresponds to the value of the pixel, that is to say, the spatial average over 10×10 km². But if the moving speed is four times faster, then the time average is that of the average of four pixels, and no longer one pixel. It is not the same anymore! How many pixels should be taken? One or four? How to take into account the wind speed and its direction, or even the vertical profile of the wind speed vector? In this specific example, it does not really matter because the pixels are

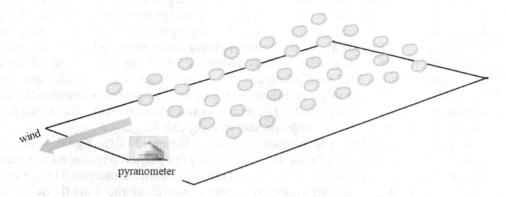

Figure 8.2 Schematic view of an infinite set of identical, small clouds, whose positions relative to each other are frozen, moving above a pyranometer at a constant speed.

expected to all have the same value. This is because clouds sizes and their spacings are small relative to the pixel size and there is approximately the same number of clouds in each pixel.

Despite its simplicity, this example illustrates the difficulty of expressing the equivalence between time and space. I do not know of a way to answer for sure a priori. An approximation of the equivalence may be obtained a posteriori by comparing a time series acquired by an instrument at ground and the time series of pixels containing and neighboring the location of the instrument. For example, one can look for the number of contiguous pixels, a kind of super-pixel, for which the correlation between the time series of the instrument and the series averaged over the pixels is the greatest. The side of this super-pixel will be equivalent on average to the duration of an observation by the instrument. For solar radiation at ground, one would typically expect a measurement lasting 1 h to correspond to a 10- to 30-km side. However, other results are possible depending on the variability of the radiation over time and space.

8.3 Fluctuations around the average – small-scale variability

A time series or spatial series of meteorological measurements generally comprises many scales. Among these are the scales that are sources of energy, generally the large scales, and the dissipative scales of energy, generally the smaller scales. Between the two sets of scales, that is to say, excluding the sources and sinks of energy, the intermediate regime can be generically called turbulence, or turbulent regime. For example, microturbulence may designate the phenomena of scales of 1 mm–100 m, or 1 ms–1 min, and macroturbulence those of 1–100 km, or 1 min–1 h. The turbulent regime can be thought of as a juxtaposition of several elementary signals, each of them having a well-defined scale (period, wavelength). The imbrication of these elementary signals gives a random structure that is difficult to predict accurately.

Let me take the example of the irradiance E_{day} received at ground on a daily average to illustrate the presence of many scales. Since E_{day} is the product of the extraterrestrial radiation E_{0day} by the clearness index K_{Tday}, the scales present in E_{day} are those present in E_{0day} and K_{Tday}. As seen in Chapter 3, E_{0day} depends only on the latitude and the solar declination. This corresponds to timescales ranging from 1 day to 1 year and spatial scales from 10 to 10,000 km. All these variations are not random and can be calculated accurately. On the opposite, the clearness index depends on the constituents of the atmosphere which exhibit a random nature. The characteristic scales of K_{Tday} are those of clouds, aerosols, and other variations in the optical properties of the cloudless atmosphere and offer a great variety. The temporal and spatial scales of E_{day} result from the juxtaposition of all these scales present in E_{0day} and K_{Tday}.

In turbulence studies, it is common to subtract the mean of the signal, which is a signal of the lowest possible frequency, and to study the fluctuations around this mean. The average of these fluctuations is zero, and this may facilitate analysis. The same can be done for irradiance or solar irradiation: remove the average and study the deviations from this average. In the previous example of the daily average irradiance E_{day}, E_{day} can be divided by E_{0day} and the study may be limited to K_{Tday}. By then subtracting the mean of K_{Tday} from the series of values, the random nature of the irradiance and its variability are identified more precisely.

There are many mathematical tools dedicated to the analysis of the temporal and spatial variability of meteorological and other variables. They generally apply to fluctuations around the mean. The best known are likely the autocovariance function and its equivalent in spectral analysis: the power spectral density, calculated using the Fourier transform.

These tools generally assume that the fluctuations form a stationary process in a weak sense. This weak-sense stationarity is also called wide-sense stationarity or second-order stationarity because its definition relates only to the first two moments of the random variable. Let $y(x) = \{y_1, y_2, y_3 \ldots\}$ be a time or spatial series of measurements. The process $y(x)$ is second-order stationary if it fulfills three conditions. First of all, its average does not vary over time. If $E[]$ denotes the mathematical expectation operator (mean), then $m = E[y(x)]$ is a constant and does not depend on the first element chosen in the series to perform the calculation. In other words, given a sufficiently long series of values, the averages that can be calculated on parts of the series, themselves sufficiently long, must be identical to each other, as well as to the average of the series. Obviously, a series comprising a temporal or spatial drift, that is to say, a constant increase or decrease in values, is not stationary. The second condition is that the variance $var(y)$ of $y(x)$ exists; that is, it is finite and is independent of time or space. The third condition is that the autocovariance function $K(x, d)$ is invariant by translation; i.e., it depends only on the shift d, that is, $K(x, d) = K(d)$.

There are many processes in nature that are not stationary, not even stationary in a weak sense. Solar irradiance received at ground is one of these processes. Suppose a series of measurements of irradiance E taken every 1 min for 6 months. During these months, the extraterrestrial irradiance E_0 varied depending on the sun–Earth geometry, inducing an underlying variation in the mean and variance of E over the months, thus violating the conditions for a weak-sense stationarity. The same goes for space. Let me refer to Figure 6.9 (Chapter 6) that presents a map of the annual average of the irradiance E received at ground for the period 1990–2004. This map shows a partially zonal distribution of E. This variation of E with latitude induces a spatial trend, which means that the mean and the variance are not constant over a territory larger than a few tens of km, thus violating the conditions for a second-order stationarity.

Non-stationarity can be induced by large-scale variations of the signal, even after removing the mean and looking only at fluctuations. Do increments between successive values, or more distant, respect the weak-sense stationarity? Indeed, suppose a line like $y(x) = a\, x$. The process $y(x)$ is not weak-sense stationary because it continuously increases with x. However, the increments $(y(x+d) - y(x))$ are equal to $(a\, d)$ and do not depend on x; they only depend on the gap d. The variance of the increments is also finite and independent of the first element chosen in the series to perform the calculation. The autocovariance function $K(x, d)$ is invariant by translation. Therefore, the process $(y(x+d) - y(x))$ is weak-sense stationary.

The process $y(x)$ verifies the weak-sense stationarity of the increments if for all d, the increments $(y(x+d) - y(x))$ have an expectation and a variance independent of the first element chosen in the series to perform the calculation. The expectation $m(d)$ and the variance var(d) of the increments are given by:

$$m(d) = E\left[y(x+d) - y(x)\right] \tag{8.1}$$

$$\mathrm{var}(d) = E\left[\left(y(x+d) - y(x)\right)^2 - m^2\right]$$

The variogram $\gamma(d)$ is often used in the analysis of variability. It describes the way in which the increases evolve on average as a function of the gap d. It is defined as follows:

$$\gamma(d) = (1/2)E\left[\left(y(x+d) - y(x)\right)^2\right] \tag{8.2}$$

Note that if $y(x)$ is itself weak-sense stationary, then the variogram is related to the autocovariance function $K(d)$ by:

$$\gamma(d) = K(0) - K(d) \Leftrightarrow K(d) = K(0) - \gamma(d) \tag{8.3}$$

Generally, there is no longer any correlation between the measurements when d is large: The autocovariance function $K(d)$ decreases and tends toward 0. Then, $\gamma(d) = K(0)$ when $d \rightarrow \infty$, and it comes:

$$K(d) = \gamma(\infty) - \gamma(d) \tag{8.4}$$

To illustrate the behavior of the variogram and its interest in the analysis of variability, let me take the example of a very dense network of pyranometers, as mentioned at the beginning of this chapter. The pyranometers are ideally identical with low measurement noise. Figure 8.3 illustrates such a network with pyranometers regularly spaced 2 km apart, over a distance of 100 km in both north–south and east–west directions. If n is the number of pyranometers on a horizontal line of the figure, or vertical since the network is symmetrical, then there are n^2 pyranometers, with, here, $n = 50$. The first pyranometer is located at point x_1, the second at x_2, etc., up to point x_{n^2}.

Let me suppose that this network is located in a homogeneous zone vis-à-vis the radiation, i.e., without great major fluctuation in space. This network is ideal for

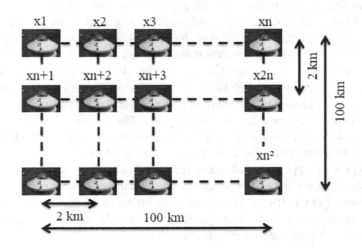

Figure 8.3 Illustration of a dense and regular network of identical pyranometers, spaced 2 km apart over an area of 100 × 100 km².

studying fluctuations in radiation at small scales, from 2 to a few tens of km. Pyranometers are very close to each other, and the hourly irradiation is expected to be quite similar from pyranometer to pyranometer, at least over some distance. In fact, the few experiments made with dense networks, quite similar to our ideal network, show significant variability, even for fairly small distances d between instruments. The variogram clearly shows this variability.

Let $y(x_1)$ be the measurement for the pyranometer located at x_1 in the network in Figure 8.3. A first series of increments is obtained by calculating the difference between this measurement and the measurements of the pyranometers located at a distance d of 2 km: $\left\{\left(y(x_2)-y(x_1)\right),\left(y(x_{n+1})-y(x_1)\right)\right\}$. By repeating for the other pyranometers located at x_2, x_3, etc., a complete series $\left(y(x+d)-y(x)\right)$ is obtained for the increment $d=2$ km and $\gamma(d=2)$ can be calculated. The operation is then repeated, but this time, setting $d=4$ km, then 6 km, and so on, up to $d=100$ km. The variogram $\gamma(d)$ is thus obtained whose typical shape is that given in Figure 8.4. Here, the variogram is represented as a percentage relative to the square of the mean of the irradiation.

If the variability had been zero, i.e., if the hourly irradiation had been the same at all pyranometers, excluding the measurement noise, then the variogram would have been a flat zero curve. Indeed, the differences $\left(y(x+d)-y(x)\right)$ would have been zero for all x and all d, and therefore, the variogram would have been zero. Figure 8.4 shows that this is not the case: The radiation exhibits a significant natural variability, even for small distances d between sites.

By definition, the function $\gamma(d)$ is positive. $\gamma(d)$ generally increases with increasing d because the differences in irradiation, in absolute terms, increase with d. This is why the curve in Figure 8.4 increases as the distance d increases. The variogram grows rapidly at first; then, the growth declines markedly for $d > 50$ km and tends to reach a kind of plateau. A plateau means that the variogram no longer increases as the distance between sites increases and the spatial correlation between the increments

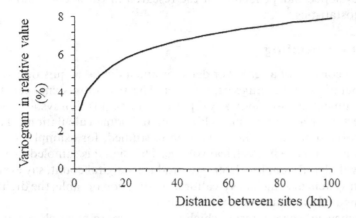

Figure 8.4 Example of variogram of hourly irradiation as a function of the distance, expressed relative to the square of the mean of the irradiation.

becomes zero. In practice, this plateau located at about 100 km means that the fluctuations in the irradiations measured at 100 km away are not correlated.

By definition, $\gamma(d)$ is equal to half of the mean square of the increments $E\left[(y(x+d)-y(x))^2\right]$ for given d. Therefore, an estimate of the mean magnitude of the increments is obtained by taking the square root of twice $\gamma(d)$. On this curve, $\gamma(d)$ is about 4 % of the square of the average hourly irradiation for $d=10$ km. Knowing that the square root of 0.08 (two times 4 %) is 0.28, this means that on average, the increments for $d=10$ km are about 28 % of the average irradiation. This relative value is important; in a practical way, it means that measurements 10 km away show differences, the relative amplitude of which is of the order of 28 % of the average irradiation. These differences are amplified when d increases: The amplitude is of the order of 40 % for distances of 100 km. It is concluded from this illustration that the variability of solar radiation at small scales is significant and should not be neglected, even if it is difficult to estimate.

The variogram shown here in Figure 8.4 is only an example, but it is representative of variograms computed with hourly irradiation measurements. The fluctuations in space are greater with measurements taken every 1 min. The spatial variogram then has greater amplitude and a plateau reached at a smaller distance compared to Figure 8.4. On the contrary, a spatial variogram calculated with daily values shows lower amplitude and a plateau reached at a greater distance compared to Figure 8.4. The variogram was calculated here with a spatial network of measurements to quantify the variability in space. It can also be calculated with a time series of measurements, the distance d then being replaced by a time shift.

The variogram is equal to the structure function, often noted $D(d)$, except for a factor 2: $2\,\gamma(d)=D(d)$. The structure function was one of the preferred tools of the researchers in the ex-Union of the Soviet Socialist Republics to model the turbulence in the atmosphere and more generally in fluids, while their counterparts in Western Europe and America were using autocovariance functions and power spectral density.

The variogram was developed independently of the research in turbulence and is a founding tool of the geostatistics.[2]

8.4 Measurement – sampling

An instrument typically consists of a sensor or detector and a digital acquisition system with digital computation and storage capabilities. The two subsystems can be integrated into the instrument, or else physically separate. The acquisition system controls and monitors the sensor and performs calibration and some calculations. The sensor output signal corresponding to the phenomenon studied, for example, solar irradiance, is generally a difference in electrical voltage. This signal is sampled by the acquisition system – I will detail sampling later – to produce a digital count, say every 1 min. If the acquisition system also includes a calibration function or table, the digital count can be transformed into a measurement of irradiance.

Measurement is a somewhat general term, which is defined more precisely according to the instrument and its use. In the example above, the measurement can be a digital count. Once calibrated, it is irradiance. If the instrument also contains signal correction functions based on given environmental variables, or even downloaded at each time of measurement, a corrected measurement of irradiance is obtained. These three scenarios show that measurement is an intuitive notion, of course, but sufficiently fuzzy that it has to be clearly specified for each use.

In general, whatever the variable to be measured, the instrument samples this variable; i.e., it collects at defined, generally regular intervals the values of the observed signal integrated over these intervals. This operation is called sampling. The discrete values are sometimes called samples. Sampling is done in time, or in space in the case of an image, or even time and space simultaneously for a series of images.

Things are a little more complex when the instrument is equipped with computation and digital storage capabilities, which very often is the case. The instrument samples itself the detector signal with a regular step, for example, 1 min. This sampling, which I can call internal, is generally defined by the instrument manufacturer, and the user has no control over it. The corresponding measurements, which I call here intermediate, are stored temporarily in the instrument or the acquisition unit. The latter generally offers several possibilities to the user. Either the intermediate measurements are delivered to the user, or the on-board intelligence performs an average or a sum of several successive intermediate measurements during a certain period, for example, 1 h or 1 day, and only delivers these results and not intermediate measurements. This operation controlled by the user actually corresponds to a second sampling. This second sampling is known to the user and controllable unlike the first.

In the following, for the sake of simplicity, I forget about the first sampling and I only consider the one controlled by the user. Actually, it is the result of the second sampling controlled by the user that is a measurement from the user's point of view. Integration over a period of time, for example, 1 h or 1 day, is called duration of integration or integration duration or integration period or summarization. This notion of duration

---

2 Geostatistics is a branch of statistics dealing with spatialized and time-varying data. It was initiated in the year 1960 by Georges Matheron, from the École des Mines de Paris, currently Mines ParisTech.

of integration can be generalized to the results of summation or averaging procedures applied to measurements. A daily irradiation can be obtained by the instrument, or by summing up 24 hourly irradiations. In this case, it is said that the integration period is 1 day.

In the case of an image, the integration is spatial and performed on one pixel. An imager like those on board Earth observation satellites is characterized, among other things, by its instantaneous field of observation, which is, in a way, the aperture angle of the detector at each instant of observation. This angle defines the pixel that is the ground surface from which the observed signal originates. Other instruments such as vertical sounders on board satellites are not strictly speaking imagers but also have an instantaneous field of observation, which defines a pixel, which it is customary to call a cell in the field of study of these sounders. This notion of pixel can be generalized to the outputs of numerical meteorological or climatic models. In these models, calculations are performed on grids, often at nodes, i.e., at intersections of the grid, but sometimes in every cell of the grid. The outputs are equivalent to measurements representing areas equivalent to the size of the cell. These models are three-dimensional; that is, they take into account the vertical dimension. It is therefore sometimes more appropriate to speak of voxels. In the case of the radiation received at ground, it is sufficient to deal only with the surface layer of the model, which is considered two-dimensional and made up of pixels.

The pixel is an integration surface and is the analogue of the integration duration. Researchers in information science use the term information support, a generic term which designates the surface area, or even the volume, as well as the integration duration, or both together.

The time interval or the distance separating the beginnings or ends of two consecutive measurements is called the sampling step. Its dimension is time or length depending on the case. In the case of a sampling in time, the inverse of the sampling step is called sampling frequency. It is often expressed in hertz (Hz, with $1\,\mathrm{Hz} = 1\,\mathrm{s}^{-1}$). The number of samples per unit of time is called the sampling rate. Let me take two examples in which the radiation measurements are integrated over 1 min. If the acquisition is made every minute, the sampling step is 1 min, the sampling frequency is 0.017 Hz ($= 1/(60\,\mathrm{s})$), and the sampling rate is 60 samples per hour. If the acquisition is made every 10 min, i.e., if one measurement lasting 1 min is made every 10 min, the sampling step is 10 min, the sampling frequency is 0.0017 Hz ($= 1/(600\,\mathrm{s})$), and the sampling rate is six samples per hour.

A sampling is regular if the measurements, or samples, are acquired over the same information support and follow each other without interruption. If the two conditions are not met, the sampling is said irregular. In the two previous examples, the sampling is regular in the first example (1 min) and irregular in the second example (10 min).

Sampling is often regular for the outputs of numerical meteorological, climatic, or other models. Outputs are contiguous in time and space. The calculation step is equivalent to a temporal information support and is regular. If the cell size is the same for all cells, the sampling is also regular in geographic space. In some numerical models, the size may depend on the geographic position of the cell and the sampling is irregular in geographic space.

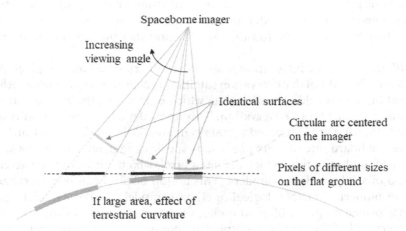

Spaceborne imager

Increasing
viewing angle

Identical surfaces

Circular arc centered
on the imager

Pixels of different sizes
on the flat ground

If large area, effect of
terrestrial curvature

Figure 8.5 Illustration of sampling by a spaceborne imager.

The sampling by a spaceborne imager is slightly more complex as shown in Figure 8.5. The instantaneous field of view is constant. Thus, sampling is regular along an arc centered on the imager and the observed surfaces (information supports) are identical (Figure 8.5). When projected onto the ground, the pixels do not have the same size for various reasons. For example, a pixel located far from the one under the satellite is seen with a greater viewing angle, which mechanically results in a larger surface (Figure 8.5). The curvature of the Earth accentuates the effect of the viewing angle if the pixel is more than 100 km away from the vertical of the satellite (Figure 8.5). The relief is another cause of the discrepancy in pixel size. Therefore, the sampling corresponding to the projected observation on the ground is not entirely regular: The pixels are consecutive, but their size on the ground is variable. Finally, these spaceborne instruments do not offer regular temporal sampling. A pixel is seen in approximately 0.1 ms (temporal support of 0.1 ms), but only every 10 or 15 min for geostationary satellites and more irregularly for other satellites.

It is possible to find ground-based instruments or sets of instruments for which the information supports are not contiguous in time or space. In other words, the sampling is not regular. For example, an instrument can deliver an integrated measurement over, say, 1 min, but only every 10 min. This choice may result from a digital storage constraint or from limited capacities in transmission of measurements. A network of measuring stations is another example as it generally does not offer contiguous spatial supports because of the distance between the stations.

Another illustration of irregular sampling is given by some ground instruments having a sight port with a spectral filter wheel such as filters in blue, green, and red. The measurements are made successively by rotation of the wheel: first a measurement in the blue range, then in the green, and finally in the red, and the cycle starts again. The time the wheel rotates until the corresponding filter is in place and then stabilizes is an interval during which there is no measurement. The time of a complete rotation is the time required to have two consecutive measurements in the same spectral band, and the temporal supports of the measurements are therefore non-contiguous.

8.5 Sampling theorem – detection of the smallest details

Reader, you understand that the smaller the sampling step, the better the observation of small-scale variations in signal and the better the perception of fine details of the signal. With a sampling interval of 1 h, it is possible to clearly distinguish the variations in radiation during a day. On the other hand, it is impossible to analyze the intra-hourly variations. You could tell me: Why not sample systematically with a very fine step? Yes, of course, but high-frequency sampling requires a large amount of data to be stored and transmitted. A trade-off must be found between your wishes, the constraints of the instruments, and the storage and computing capacities.

8.5.1 Sampling theorem

You must first determine the size of the smallest details you want to analyze in time or space. The inverse of this smallest size of interest defines a frequency, called the maximum frequency since it is the largest you want to see. The optimal sampling frequency is then given by the sampling theorem. The latter is also said to be Shannon or Nyquist–Shannon theorem. The theorem claims that the discrete representation of a signal requires regular sampling, with a sampling frequency greater than twice the maximum frequency. In other words, given the smallest size of interest, the sampling step must be less than half that size.

Let me take an example to better understand this theorem. Suppose a periodic signal in time like a sine function whose period is 1 h. This signal always gives the same value every hour. If I set the sampling step to 1 h, I always get the same value, so I cannot highlight this 1-h period. If the sampling step is 30 min, then I get two different values, which repeat at every hourly interval. The hourly repetition of this difference highlights the 1-h period. I can hold the same reasoning in the space field. If I want to see variations in size greater than 10 km, then I need a pixel that is twice as small as 5 km. If the size of the pixel is 10 km, it is impossible to highlight a phenomenon whose size of the variations is less than or equal to 10 km.

Figure 8.6 illustrates the influence of the sampling step for a time signal. The graph on the left represents a sine wave with a period of 20 arbitrary units of time. By construction, a pattern is observed that repeats every 20 units. For example, the curve takes the value 0 with an ascending slope at time steps 20, 40, 60, etc. This sine curve is the signal to sample in a regularly manner. The sampler, i.e., the sampling process, works as follows: For a given sampling step, the signal is averaged during this time interval and the average is provided by the sampler at the end of the interval.

For example, for a sampling step of 20, the original signal is averaged over 20 time units and the average is given at time 20, 40, 60, etc. This result is plotted as a dotted curve on the graph on the right (Figure 8.6). Since the mean of a sinusoid over its period is zero, the sampling result is 0 at each of these time steps 20, 40, 60, etc. There is no variation in this result, and the resulting curve (dotted line) is flat and equal to zero. Therefore, the 20-unit period cannot be detected.

The sampling theorem says that the sampling step should be at most equal to half of 20, i.e., 10. The result of sampling at step 10 is represented by the solid black curve in Figure 8.6. This time, a pattern can be clearly seen that repeats every 20 units. For example, the black curve takes the value 0 with an ascending slope at time steps

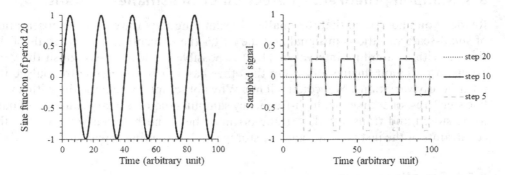

Figure 8.6 Illustration of the influence of the sampling step. (a) The signal to sample regularly, which is a sinusoid whose period is 20 arbitrary units. (b) The results of the sampling with a step of respectively 20, 10, and 5 arbitrary units.

20, 40, 60, etc., like the original curve. As expected, the period of 20 units is clearly highlighted with this sampling. If the period is found, one notes on the other hand that the amplitude of the result does not correspond to that of the original signal. The latter is equal to 2, from −1 to +1, on the left graph, whereas it is worth approximately 0.6, varying from −0.3 to +0.3, for the signal sampled with a time step of 10. This illustrates that the variability of the signal and therefore of the radiation at scales less than twice the sampling interval cannot be reproduced correctly. The size of the details (here, the period) is detectable, but not their amplitude.

The graph on the right also represents the result of sampling with a step of 5 units, which is twice the minimum frequency required by the theorem. The gray dashed curve shows a pattern that repeats every 20 units and clearly shows the existence of a 20-unit period. The amplitude is around 1.8, varying from −0.9 to +0.9, and is quite close to the original amplitude.

In practice, if the theorem indicates that a sampling step equal to half of the smallest period sought, or more generally of the smallest size of the details of interest, is sufficient to detect these details, it is better to sample a bit more frequently. One can take, for example, a step of a third of the smallest size of interest. This makes it easier to describe the smallest details, especially in the presence of noticeable measurement noise.

8.5.2 What are the smallest observable details in a series of measurements?

The sampling theorem indicates the optimal sampling frequency. Conversely, this theorem gives the maximum detectable frequency in a series of measurements already acquired, or in an equivalent manner, the size of the smallest detectable details. This is an important result when you are using measurements acquired by third parties and you have not chosen the sampling step yourself.

If I use the previous examples, I can detect periods of 1 h and more, by having a series of measurements acquired with a regular sampling whose step is 30 min.

Similarly, I can detect wavelengths of 10 km and more, on an image acquired with a regular sampling whose step is 5 km. If the outputs of a meteorological model are hourly and with square pixels of 50 km in size, I can detect, in a series of outputs, periods of 2 h and greater and wavelengths of 100 km and greater. I write here period and wavelength for reasons of convenience. It is necessary to understand details, or variations, detectable in time and space.

The case of irregular sampling is somewhat more complex. Let me first take the case of the ground instrument with a filter wheel, discussed above. Suppose that the acquisition of the signal for a given filter is done in 10 s and that the period of rotation of the wheel is 1 min. Then, the smallest detectable period is twice the rotation period of the wheel, i.e., 2 min.

A slightly more difficult case is that of the spaceborne imager mentioned above, with a temporal support of 0.1 ms, but sampled in time every 10 min, and with a spatial sampling of 5 km, which is assumed to be regular for reasons of simplification. Since the spatial sampling is regular, the smallest detectable detail has a size of 10 km (twice 5 km). But what about time? What is the smallest detail detectable in time? This is not a priori a simple question to solve from a mathematical point of view, as seen earlier in the discussion on the equivalence between space and time. Thanks to the ergodicity hypothesis, the spatial integration of the signal on the pixel is equivalent to integration over time. The signal acquired for 0.1 ms on the pixel of 5 km in size can be considered as representative of a sample lasting a few minutes. Since the sampling is regular in space, the spatial series can be considered equivalent to a time series with a more or less regular sampling step of a few minutes. This regular pseudo-sampling in space counterbalances the irregular temporal sampling of 10 min. It can then be estimated that the smallest detectable period is around 20 min as a first approximation.

Chapter 9

Ground-based instruments for measuring solar radiation at ground

---

At a glance

Earth observation is the gathering of information about the physical, chemical, and biological systems on Earth, whether natural or anthropogenic, as well as the means of acquiring, evaluating, and modeling this information and systems. The estimation of the solar radiation received on the ground, total or spectral, thus calls upon all the means of information acquisition: instruments on the ground or on board of sounding balloons, planes or satellites, and more or less complex numerical models.

The best-known instruments located at ground are pyranometers, pyrheliometers, and pyranometers equipped with a shadow ring or shadow ball, which respectively measure the total radiation and its components. Radiation in specific spectral bands is measured using specific detectors or spectral filters. Uncertainties in measuring radiation depend on the instruments, their implementation and daily maintenance, and the monitoring of their calibration. Three categories of measurement quality are distinguished by the World Meteorological Organization: high, good, and medium, also called moderate. High-quality measurements present the smallest uncertainties; this quality is very difficult to obtain.

Whether for total or spectral radiation, the networks of ground-based stations do not cover the surface of the Earth equally. There are often too few stations to describe the geographic distribution of radiation. Often, the series of measurements are incomplete and contain gaps, compromising the temporal analysis of the radiation. Temporal or spatial averages must take these gaps into account. A simple approach to the completion of a series consists in interpolating and extrapolating the existing values, taking into account the variability of the radiation and possibly using the clearness index, or the clear-sky index.

The measurement and estimation of solar radiation received at ground at a given location at a given time can be done in several ways, whether for total radiation or radiation in a spectral range such as ultraviolet (UV), or daylight, or photosynthetically active radiation (PAR). The present chapter and the following intend to briefly describe the most common ways.

Earth observation is the gathering of information about the physical, chemical, and biological systems on Earth, whether natural or anthropogenic, as well as the means of acquiring, evaluating, and modeling this information and systems. The estimation of the solar radiation received on the ground, total or spectral, thus calls upon all the means of information acquisition: instruments on the ground or on board of sounding balloons, planes or satellites, and more or less complex numerical models. These means can be combined using interoperability techniques, developed in particular under the influence of Group on Earth Observations (GEO),[1] which makes solar radiation estimation a very vast, rich, and abundant field of study. Improvements are regularly made to observation and measurement, particularly with regard to miniaturization, on-board electronics and digital processing, communications between sensors, and mathematical techniques for collaboration between them (data fusion[2]). There are also advances in the numerical modeling of radiative transfer in the atmosphere. Thanks to this, measuring instruments on the ground or on board satellites and numerical models evolve rather quickly as well as their possible combinations, for a more accurate estimation of the solar radiation received at ground and of its variations in time and in geographic space.

It is not my ambition to draw up an exhaustive range of all the instruments and models or even to draw up a diagram of the avenues currently being explored by researchers. That would be presumptuous on my part and probably unnecessary. Indeed, I have spent enough years doing research and development on this subject to know that a little imagination, combined with a good knowledge of the means implemented, often opens the way to a new approach, a new relationship, or a new instrument. This is why these two chapters are rather short compared to the large number of scientific publications available. I just briefly describe the main instruments used at the time of writing the book and the most common ways to estimate solar radiation received at ground. The present chapter deals with the direct means for measuring the radiation at ground and presents the most common instruments for measuring radiation at ground. The indirect means are discussed in the following chapter.

This chapter begins with the presentation of the most widely used detectors in instruments for measuring solar radiation received at ground. Next, I give the uncertainties associated with the radiation measurements as recommended by the World Meteorological Organization.

Ground-based measuring stations, whether for total or spectral radiation, are generally too few to accurately describe the geographic distribution of solar radiation. The lack of financial and human resources for the installation and especially the maintenance of a station and even more of a network is the main cause of the scarcity of stations. I illustrate this lack of stations with a few examples taken from the solar radiation database of the World Meteorological Organization. In addition, the time series of measurements are often incomplete and have gaps, compromising the temporal

---

1 GEO is a partnership of more than 100 national governments and in excess of 100 participating organizations which envisions a future where decisions and actions for the benefit of humankind are informed by coordinated, comprehensive, and sustained Earth observations (from its Web site http://www.earthobservations.org).

2 See, for example, Wald L., 2002. *Data Fusion. Definitions and Architectures – Fusion of Images of Different Spatial Resolutions*, Presses de l'Ecole, Ecole des Mines de Paris, Paris, France.

analysis of the solar radiation. It is rare to have sufficiently complete series of measurements to allow an accurate description of the solar radiation received on the ground.

It is common to calculate the average of a time series of radiation measurements, for example, the annual average of irradiance. This is a simple operation that is described at the end of this chapter. However, it can be made difficult because of gaps in measurements in the series. Temporal or spatial averages must take these gaps into account. How to do it? Asking this question naturally leads to discussing the completion of an incomplete series of measurements.

9.1 Common ground-based instruments

On several occasions, I have already mentioned in this book the pyranometer, the instrument commonly used in meteorological stations and others, for the measurement of the solar radiation received at ground. This section provides an opportunity to describe pyranometers in more detail. Other commonly used instruments are also presented. However, this section does not aim to list all the instruments used in the world or the different developments of new sensors and instruments. This is a very active area of research and development, and it would be presumptuous of me to pretend to cover it.

As I mentioned before, an instrument is made up of one or more sensors and includes on-board intelligence, storage, and communication capabilities. It can possibly use remote computing resources or external knowledge. Technological advances in miniaturization, electronics, and on-board digital processing offer many opportunities for improvement. Thanks to them, it is relatively easy to invent new sensors and instruments or new ways of exploiting existing sensors. Collaborative sensors and instruments are instruments providing measurements which are themselves the fusion of several intermediate measurements and possibly calling on external knowledge and models. For example, scientists have been verifying and correcting the calibration of spaceborne sensors for several years using measurements from other spaceborne sensors. Could this example not be transposed to measurements of solar radiation at ground? To go a bit further, can a process be invented where the measurements acquired by remote low-cost sensors would feed a model predicting the measurements taken at another point by another instrument? Deviations between predictions and actual measurements, their fluctuations, and statistical distributions could provide information on the quality of the instrument calibration. Reader, my intention is not to get you lost in a discussion on sensor calibration. Through this brief discussion on a very specific point, I am only trying to show you how easy it is to innovate and, in general, to make you perceive the abundance of ideas in the development of instruments.

9.1.1 Thermopiles and photodiodes – calibration and classes of instruments

The most common instruments use two types of detectors: either thermopiles or photodiodes.

A thermopile instrument is a kind of short cylindrical tube at the bottom of which is a black absorbent disk. The temperature of the disk is related to the irradiance received. The greater the irradiance, the greater the temperature of the disk. The temperature difference between the disk and a so-called reference surface, not receiving any radiation, is converted into electric voltage through the phenomenon known as the thermoelectric effect. The datalogger or digital acquisition system converts voltage into irradiance through calibration (typically 10 mV for 1000 W m^{-2}). Thermopiles react homogeneously over a very wide spectral range since the measurement results from the heating of a black body. Therefore, they can be used for total radiation measurements, i.e., between 200 and 4000 nm. They can also be used for spectral radiation measurements by using spectral filters placed at the inlet of the instrument.

A photodiode produces an electric current, the intensity of which is a function of the irradiance it receives. This current is then converted into an electrical voltage, itself converted into irradiance thanks to a calibration. Unlike the thermopile, each type of photodiode has a very precise spectral response. By properly choosing the photodiode, the spectral interval to be measured can be selected. Photodiode instruments are well suited to intervals between approximately 350 and 1000 nm. A conversion model must be established if measurements from a photodiode are to be used for estimating total radiation, taking into account that the relationship between total radiation and that in a particular spectral band is a function of atmospheric conditions, as seen in Chapter 7.

The photodiode is not the only detector for spectral irradiance. Other instruments and detectors exist, such as spectrometers for measuring the spectrum, often equipped with diffraction gratings and charge-coupled device (CCD)-type detectors. Ultraviolet radiation is often measured by a spectrophotometer comprising a number of diffraction gratings and a photomultiplier tube to count the photons received.

In meteorology, all instruments comply with calibration standards based on an absolute reference: the World Radiometric Reference (WRR) of the World Meteorological Organization. The WRR is called the primary standard. Any instrument is referenced against this standard, according to one of the following three classes. The most accurate class is called the secondary standard; it is the closest to the primary standard. The other two classes are called first and second classes. First-class instruments are more accurate than second-class instruments. The differences between these two classes come from differences in uncertainties, response times, sensitivities, zero shifts, directional errors, stability, etc.

9.1.2 Pyranometer – pyrheliometer – pyranometer with a shadow ring or ball

The principle of the pyranometer is the same regardless of the spectral measurement window and whatever the type of detector: thermopile or photodiode. The instrument includes a short tube at the end of which is the detector. The opening of the tube is protected from all kinds of weather and from internal condensation by a double dome, as shown in Figure 9.1. The domes are made so as to offer the same spectral transmission regardless of the wavelength and the position of the sun. The pyranometer receives the solar radiation from the upper hemisphere and therefore provides a measurement of the global radiation.

The material used for the domes defines the spectral transmission window. The spectral response of the instrument is the product of the spectral response of the detector and the spectral transmission of the domes. Thermopile pyranometers are often used to measure total radiation. They react to radiation in a spectral interval from approximately 330 to 2200 nm, which represents about 97 %–98 % of the total radiation, as seen in the tables in Chapter 7. Figure 9.1 shows a pyranometer fitted with a sun shield

Figure 9.1 Pyranometer and its sun shield.

that is a metal truncated cone painted in white. The sun shield houses the electrical connections and a drying element to prevent condensation. It allows ventilation of the instrument and prevents unwanted heating by the sun rays of the wall of the tube housing the thermal detector.

The pyrheliometer measures the direct component of radiation at normal incidence. It aims at the sun at all times and has an opening close enough to receive only this component. The instrument essentially consists of a tube at the end of which is the detector. The tube is long so as to minimize unwanted reflections from light rays not coming from the direction of the sun. The pyrheliometer is mounted on an automatic sun tracking system. Such a system constantly follows the sun as it travels through the sky. The pyrheliometer is thus always pointed at the center of the solar disk. Figure 9.2 shows a set of instruments, for the simultaneous measurement of the direct component at normal incidence (pyrheliometer) and of the diffuse component received by a horizontal plane (pyranometer). This set is equipped with a kind of articulated parallelogram, which is the sun tracking system.

The measurement of the diffuse component on a horizontal plane is often done using a pyranometer, for which the direct component is masked by a shadow ring or a shadow ball, also called shading ball or shade ball. Figure 9.2 shows a pyranometer equipped with a shadow ball, mounted on the sun tracking system. The ball permanently masks the solar disk and removes the direct component. The shadows of the shadow ball and its arm on the pyranometer are clearly visible in the photo on the left. The shading ring has an arc shape and obscures the sun as the sun travels through the sky. The position of the ring is adjusted daily by an automated system.

Such a set made of a pyrheliometer and a shaded pyranometer is often associated with another pyranometer measuring global radiation in order to simultaneously and independently measure the global radiation and its two components. Because the global radiation is the sum of the two components, redundant measurements of the global radiation are thus obtained. This redundancy is useful to control the quality of the measurements.

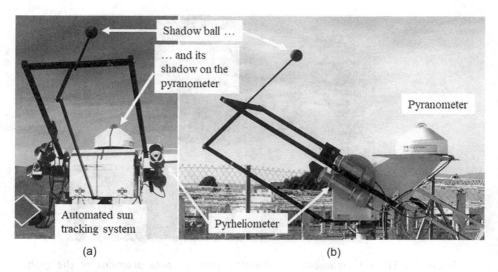

Figure 9.2 Equipment for the simultaneous measurement of the direct component of radiation at normal incidence (pyrheliometer) and the diffuse component on a horizontal plane (pyranometer). (a) Front view and (b) profile.

9.1.3 Other instruments measuring global radiation and its diffuse component simultaneously

Some instruments make it possible to measure simultaneously, or almost simultaneously, the global radiation and its diffuse component on a horizontal plane. The direct component on the horizontal plane is deduced from the two measurements by subtracting the diffuse component from the global radiation. The two types of instruments presented below do not require a sun tracking system or a shading ring or ball. They are therefore easier to install and maintain than shaded pyranometers for measuring the diffuse component. I present them because they are quite common as of this writing.

The instrument presented in Figure 9.3 and commercially known as SPN1 has no moving part. It is equipped with seven thermopiles and is therefore well suited for measuring total radiation. A mask presenting a complex geometry under the dome is clearly visible in Figure 9.3. It is designed to always obscure the direction of the sun for at least one thermopile. The obscured thermopiles thus give a measurement of the diffuse component for the part of the sky seen. The other thermopiles provide the measurement of global radiation for the other part of the sky. Using hypotheses on the distribution of radiances on the celestial vault, the on-board intelligence of the instrument reconstructs the global radiation and its diffuse component on a horizontal plane. The direct component on a horizontal plane is then deduced from the other two.

A rotating shadowband irradiometer measures almost simultaneously the global radiation and its diffuse component on a horizontal plane. Figure 9.4 shows three similar irradiometers, equipped with a movable arm in the form of an arc of a circle. The measuring cell includes the photodiode and must be horizontal. The instrument takes a global radiation measurement approximately every second. The mobile arm is positioned so as not to shade the cell as seen for the two instruments on the left in Figure 9.4. Once per minute, the arm rotates for a few seconds as seen for the rightmost instrument in Figure 9.4. The arm temporarily hides the sun during this movement, and the

Figure 9.3 SPN1 instrument for the simultaneous measurement of the global total radiation and its diffuse component on the horizontal plane. A cover is visible under the dome. It obscures the direction of the sun for at least one thermopile at any instant.

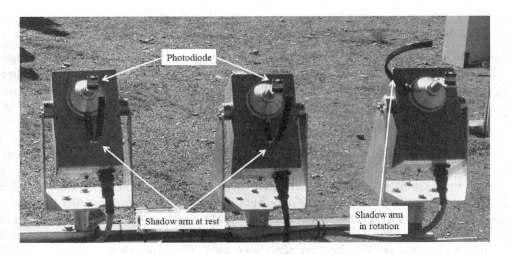

Figure 9.4 Rotating shadowband irradiometers. The arm rotates for a few seconds every minute. During the rotation, it casts a shadow on the photodiode and temporarily hides the sun.

cell measures only the diffuse component during this brief moment. It is then assumed that the diffuse component is constant for 1 min. Each minute, the instrument gives the global irradiation and its diffuse component on the horizontal plane, the direct component being deduced from the other two. As already said, rotating shadowband irradiometers are equipped with photodiodes. They therefore have specific spectral responses depending on the photodiode chosen and its implementation. A conversion model is usually provided when the irradiometer is used for estimating total radiation.

Table 9.1 Relative uncertainties on the values measured by a pyrheliometer for three durations of integration depending on the category of quality of the measurements, according to the World Meteorological Organization

Quality	High			Good		
	1 min	*1 h*	*1 day*	*1 min*	*1 h*	*1 day*
Relative (%)	0.9	0.7	0.5	1.8	1.5	1
Minimum uncertainty on irradiation (kJ m^{-2})	0.6	21	200	1	54	400
Minimum uncertainty on irradiance[a] (W m^{-2})	10	6	2	17	15	5

[a] Minimum uncertainties on irradiance are not given in the guide. I deduced them from those on irradiation by dividing by the number of seconds included in the duration of integration.

9.2 Typical uncertainties of pyranometers and pyrheliometers

The World Meteorological Organization periodically publishes recommendations for the measurement of meteorological variables in its Guide to Meteorological Instruments and Methods of Observation.[3] This guide also gives valuable information on the use of instruments, their maintenance, and monitoring of the quality of the measurement, including in the field of radiation. It can be advantageously consulted before establishing a measuring station.

The World Meteorological Organization distinguishes three categories of quality of measurements in solar radiation: high, good, and medium, also known as moderate. High quality is the most restrictive, and measurements have the smallest uncertainties. It can only be achieved with a well-maintained secondary-standard pyranometer that is regularly calibrated against the WRR. Good quality can be achieved with a first-class pyranometer, provided it is well maintained and regularly calibrated against a secondary-standard pyranometer. Medium or moderate quality can be achieved with a well-maintained and regularly calibrated second-class pyranometer.

Table 9.2 Relative uncertainties on the daily values measured by a pyranometer depending on the category of quality of the measurements, according to the World Meteorological Organization

Quality	High	Good	Medium or moderate
Relative (%)	2	5	10
Minimum uncertainty on irradiation[a] (kJ m^{-2})	150	400	800
Minimum uncertainty on irradiance[b] (W m^{-2})	2	5	9

[a] The minimum uncertainty on irradiation is given in the guide for the good quality only. I deduced it for the two other qualities.
[b] The minimum uncertainty on irradiance is not given in the guide. I deduced it from that on irradiation by dividing by the number of seconds included in a day of 24 h.

3 Guide to Instruments and Methods of Observation. WMO-No. 8, 2018 edition, 2018, World Meteorological Organization, Geneva, Switzerland. [Vol 1 Measurement of Meteorological Variables, Chapter 7, Measurement of Radiation.]

Reader, attentive as you are, you will have noticed the emphasis on the implementation of the instruments, their maintenance, and their calibration. As seen previously, the sources of noise and systematic error affecting a measurement are numerous and accessory parts of the instrument, poorly maintained, or badly sized, or partially defective, can decrease the quality of the measurements, even if the instrument is itself almost flawless.

The statistical distribution of the random errors in the measurements is assumed to follow a Gaussian law. The uncertainty is expressed as the 95 % confidence interval, i.e., the interval of values comprising 95 % of the samples; it is equal to twice the standard deviation of the series. I report in this section the uncertainties given in the aforementioned guide for pyranometers and pyrheliometers for total radiation. Other fields of application of solar radiation may have other recommendations regarding uncertainties.

Table 9.1 reports the relative uncertainties on the direct component of the total radiation measured by a pyrheliometer for high and good measurement qualities and for three durations of integration. These values are read from the aforementioned guide. If the uncertainty on irradiation obtained by applying the relative value is less than the minimum uncertainty given in the table, the latter should be taken as the uncertainty. For example, if the uncertainty on a measurement of hourly irradiation of high quality is less than $21\,\text{kJ m}^{-2}$, the uncertainty on this measurement must be set to $21\,\text{kJ m}^{-2}$. The guide only provides minimum uncertainties on irradiation. I deduced the corresponding minimum uncertainties on irradiance by dividing those for irradiation by the number of seconds included in the duration of integration.

Table 9.2 reports the relative uncertainties on the daily values of global total radiation measured by a pyranometer for the high, good, and moderate qualities. The aforementioned guide states that measurements of daily irradiation H_{day} offered to third parties in the context of international exchanges must be at least of good quality. According to the table, the accepted uncertainty on daily measurements is equal to 5 % of H_{day} and it cannot be less than $400\,\text{kJ m}^{-2}$.

Only the minimum uncertainty on daily irradiation of good quality is given in the guide. How did I estimate the minimum uncertainties for high quality and moderate quality reported in Table 9.2? As written before, the uncertainty on daily irradiation of good quality is 5 % of H_{day} and it cannot be less than $400\,\text{kJ m}^{-2}$. From that, a limit equal to $8000\,\text{kJ m}^{-2}$ ($= (1/0.05) \times 400\,\text{kJ m}^{-2}$) can be deduced. In other words, if $H_{day} \geq 8000\,\text{kJ m}^{-2}$, the uncertainty for good quality is equal to $0.05\,H_{day}$. Otherwise, it is set to the minimum uncertainty of $400\,\text{kJ m}^{-2}$ ($0.4\,\text{MJ m}^{-2}$, or $40\,\text{J cm}^{-2}$). Starting from this, if I assume that the limit of $8000\,\text{kJ m}^{-2}$ is applicable to the high and moderate qualities, I obtain the minimum uncertainties for these two categories. The minimum uncertainty is equal to $150\,\text{kJ m}^{-2}$ ($= 0.02 \times 8000\,\text{kJ m}^{-2}$) for the high quality and to $800\,\text{kJ m}^{-2}$ ($= 0.10 \times 8000\,\text{kJ m}^{-2}$) for the moderate quality. These minima are reported in Table 9.2. Finally, dividing the minima by the number of seconds included in a day of 24 h ($86,400\,\text{s}$) yields the minimum uncertainties on daily averages of irradiance.

In theory, if the partial errors are well decorrelated and the statistical distribution of errors is Gaussian, the uncertainties on the monthly average of daily irradiation or the monthly mean of irradiance can be estimated by dividing the values in Table 9.2 by the square root of the number of days in the month. Thus, for the moderate quality, the relative uncertainty of 10 % on the daily irradiation becomes in theory about 2 % on the monthly average. The reality is more complex. Instead, uncertainty estimates published in scientific papers show a relative uncertainty of about 5 % for monthly averages and about 2 % for annual averages.

Table 9.3 reports the relative uncertainties on the hourly measurements made by a pyranometer for the high, good, and moderate qualities. The relative uncertainties are much greater than for pyrheliometers; they are also greater than for daily irradiation.

Unlike the previous cases, the guide from the World Meteorological Organization does not mention minimum uncertainty on the hourly values measured by a pyranometer. The absence of such a mention in Table 9.3 may mean for a non-expert person that the uncertainty is very low when irradiation is low, especially at the start and end of the day. However, experience shows that it is at these times that the uncertainties are greatest. I recommend that you exercise caution in using the relative uncertainties given in Table 9.3 and only consider them valid for solar zenithal angles less than 60°. It is better to multiply the relative uncertainty by two or three at greater solar zenithal angles.

In the absence of other indications, the relative uncertainties in Table 9.3 can also be adopted for intra-hourly measurements, i.e., for measurements made with an integration time of less than 1 h. To be on the safe side, they can be multiplied by a factor of 2 or 3; the shorter the integration time, the greater the factor.

The World Meteorological Organization guide deals with the measurement of spectral radiation but does not give a precise indication of the uncertainties of this type of measurement. The guide emphasizes the difficulty, on the one hand, in making accurate and high-quality measurements and, on the other hand, in estimating all the uncertainties attached to the measurement. With regard to measurements of hourly irradiations in ultraviolet, it is estimated that a relative uncertainty of 5 % at 300 nm can only be achieved under the most stringent measurement conditions. Regarding PAR, instrument manufacturers indicate that a relative uncertainty of 8 % can be obtained for hourly values, provided the measurement conditions are very good.

9.3 Incomplete coverage in both space and time by stations

Whatever the spectral measurement window (UV, PAR, total radiation, or others), the measuring stations are unevenly distributed on the surface of the Earth. They are relatively numerous in countries paying attention to solar radiation for a variety of reasons, such as protecting human health or developing solar energy, and having sufficient resources to develop and maintain these stations. On the other hand, there are very few in hard-to-reach places, such as oceans or desert areas, or in countries that do not have the necessary means and resources.

Table 9.3 Relative uncertainties on the hourly values measured by a pyranometer depending on the category of quality of the measurements, according to the World Meteorological Organization

Quality	Relative (%)
High	3
Good	8
Medium or moderate	20

This scarcity, this lack of stations, is detrimental to a good knowledge of the radiation received on the ground. I illustrate this with the example of the total radiation database[4] of the World Radiation Data Centre (WRDC), an institute of the World Meteorological Organization. This database includes measurements of radiation and duration of sunshine carried out by stations in meteorological networks collaborating in this standardized collection and archiving effort. At the time of writing this book – i.e., in 2020, – I have counted the number of stations for which the database contains time series of daily irradiation. This base also contains time series of hourly irradiation, but for a smaller number of stations.[5]

First, I counted the number of stations having measured daily irradiation for at least 1 year, between 1964 and 2019, inclusive, for each of the seven regions defined by the World Meteorological Organization. The results are shown in Table 9.4. Unsurprisingly, regions that are sparsely populated (Australia and Oceania) or difficult to access (Antarctica) offer few stations, compared to others. Europe has the largest number of stations: 528. Yet, the situation is not entirely satisfactory in Europe for this period: There are large areas with few stations. The gap between Europe and the other regions is striking. Outside of Australia and Oceania, Europe is the region with the smallest area and has the most stations by far. For example, its area is three times smaller than that of Africa and the number of stations is more than double. This difference highlights how rare stations are outside of Europe.

I then carried out a count of the stations located in the 50 or so states in Africa, excluding European dependencies. I limited the count to stations offering at least 2 years of daily irradiation measurements over the last 20 years, i.e., between 2000 and 2019. I found that only six states offered at least one station. This is a small

Table 9.4 Number of stations that measured daily total irradiation during at least 1 year, between 1964 and 2019, according to the database of the World Meteorological Organization, in each of the seven regions

Number of the region	Region	Area of the region, in millions km^2	Number of stations
I	Africa	30	235
II	Asia	45	123
III	South America	18	135
IV	North America	24	166
V	Australia and Oceania	8	87
VI	Europe	10	528
VII	Antarctic	13	10

4 World Radiation Data Centre, http://wrdc.mgo.rssi.ru/.

5 Weather information is considered strategic for the armed forces. For example, in the 20th century, a significant cloud cover over a region meant that aviation could not take off in that region or that the armies could carry out movements on the ground without these being detected by spy planes from another state. A good knowledge of weather conditions served military strategy. There was then a tension between the armies and the civilian world of meteorologists during this century as the former were eager to keep information secret while the latter were keen to share it between meteorological services around the world as local meteorology is influenced by phenomena at much larger scales.

number considering the vastness of this continent. In addition, though there are a total of 30 stations in these six states, they are unevenly distributed. There are 15 stations in Egypt, one in Algeria, one in Morocco, four in Kenya, seven in Mozambique, and two in Zimbabwe (list ordered by latitude, north to south, according to meteorological usage). It is impossible with so few stations, so unevenly distributed, to accurately describe the solar radiation incident on the ground on the African continent, in varied climates.

By way of comparison, I found more than 100 stations in the European territory of France, a state with a long tradition of measuring solar radiation. In contrast, there are only 12 stations in China, a country comparable in size to that of Europe.

This example from the World Meteorological Organization database is for illustrative purposes only. There are networks other than those of the national meteorological services, such as those installed for agro-meteorology, hydrology, or solar energy. However, the same scourge of the scarcity of measuring stations applies to all networks because the causes are the same: lack of financial and human resources for installation and especially for maintenance. As a result, if a person managed to access the measurements from all networks, she would still find loopholes in some areas.

Apart from the scarcity in space, I also want to highlight the lack of stations for long series of good quality measurements, i.e., time series spanning several years with few gaps. This remark is true for all stations around the world. This lack is greater the longer the duration of the requested measurement period. I have shown the influence of astronomical seasons on variations in solar radiation received at ground. I have also shown the great importance of the presence of clouds on these same variations. Therefore, a good knowledge of the solar radiation at ground at a point requires a sufficiently long series of measurements without gaps to capture astronomical and meteorological fluctuations. Take the case of a solar farm to be installed. It will operate for about 40 years and will experience variations of difference timescales. How to decide on the location of this farm and size it, if the measurements cover only 1 year? Is this year exceptional or close to average? And even in the case where 3 years of measurements are available, how can investors be confident in such a small number of years to decide to invest the considerable sums needed? Let me look again at the example of the African continent and the database of the World Meteorological Organization. I counted the number of stations offering fairly comprehensive series of measurements for at least 20 years, not necessarily consecutive, between 1990 and 2019 (30 years). I found only seven stations: four in Egypt, one in Algeria, one in Morocco, and one in Mozambique. This scarcity affects not only Africa, but the whole world. For example, I found only 24 stations out of the 100 stations in European France and eight in China out of the 12 stations. It is therefore rare to have sufficiently complete series of measurements to allow an accurate description of the solar radiation received at ground.

The stations are generally arranged in places consistent with the objectives of the observation. For example, stations belonging to weather networks are often located in airports or large metropolitan areas. Those of agro-meteorological networks are rather located in agricultural areas. If the objective for which measurements are desired is quite different from that of stations already installed, for example, for human health, or the development of solar electricity in sparsely populated areas, it is likely that measurements acquired by a nearby station will be rarely available. The next chapter discusses ways to alleviate this problem.

9.4 Calculation of daily, monthly, and yearly sums and averages

When a time series of radiation measurements is available, it is common practice to calculate the average of the radiation or the sum over a period. These are a priori simple operations. Actually, they are made complex by conventions in meteorology or by the possible absence of measurements in the series of measurements. Time averages must take these conventions and gaps in measurements into account. This is the subject of this section. Asking how to do it naturally leads to discussing the completion of an incomplete series of measurements.

The absence of measurements, called gaps, can be due to various factors, such as a temporary inability of the instrument, a communication fault, or a rejection of the measurement because it does not pass a quality check. A series of measurements having no gap is said to be complete. Otherwise, it is said to be incomplete. This situation very often prevails in the case of measurements made by instruments on the ground.

9.4.1 Ideal case of a complete time series of measurements

Mathematically, an average is the sum of the elements divided by the number of elements. Suppose a time series of measurements, for example, hourly irradiations measured with a sampling step of 1 h, for 24 h, from 01:00 to 24:00, knowing that the time stamp corresponds to the end of the duration of integration. The hourly irradiation is noted H_{hour}. Suppose the 24 hourly values pass the quality control satisfactorily and are declared valid, in which case the time series is complete with 24 hourly values. The daily irradiation H_{day} is the sum of these 24 values. The daily average of the hourly irradiation, noted $H_{daily\_mean\_hour}$, is equal to this sum divided by 24. The daily average of irradiance E_{day} is equal to the daily irradiation H_{day} divided by 86,400 s. For a given day j,

$$H_{day}(j) = \sum H_{hour}(t) \quad \text{for the 24 h} \tag{9.1}$$

$$H_{daily\_mean\_hour}(j) = \left[\sum H_{hour}(t) \right]\Big/ 24 = H_{day}(j)/24 \tag{9.2}$$

$$E_{day}(j) = H_{day}(j)/86,400 \tag{9.3}$$

In the following, I assume that all measurements in the series are valid; i.e., the series is complete.

The monthly irradiation $H_{month}(m)$ is the integral of the radiation over a given month m. It is equal to the sum of the daily irradiations $H_{day}(j)$ for that month:

$$H_{month}(m) = \sum H_{day}(j) \quad \text{for all days } j \text{ in the month } m \tag{9.4}$$

The monthly average of the daily irradiation $H_{monthly\_mean\_day}(m)$ is the average of all daily irradiations for that month. It can also be designated by the equivalent term: average daily irradiation for the month. If $N(m)$ is the number of days in the month m, $H_{monthly\_mean\_day}(m)$ is given by:

$$H_{monthly\_mean\_day}(m) = \left[\sum H_{day}(j)\right]\Big/ N(m) \tag{9.5}$$

The monthly mean of irradiance $E_{month}(m)$ is given by:

$$E_{month}(m) = H_{month}(m)\big/\left[86,400 N(m)\right] \tag{9.6}$$

In studies on climate change, it is common not to take into account the measurements taken on 29 February in leap years in the calculation of monthly quantities in order to better compare sums and averages of February between them.

The annual irradiation $H_{year}(m)$ is the integral of the radiation over a given year y. It is equal to the sum of the daily irradiations $H_{day}(j)$ for that year:

$$H_{year}(y) = \sum H_{day}(j) \quad \text{for all days } j \text{ in the year } y \tag{9.7}$$

The annual average of the daily irradiation $H_{annual\_mean\_day}(y)$ is the average of all daily irradiations for that year. It can also be designated by the equivalent term: average daily irradiation for the year. If $N(y)$ is the number of days in the year y, $H_{annual\_mean\_day}(y)$ is given by:

$$H_{annual\_mean\_day}(y) = \left[\sum H_{day}(j)\right]\Big/ N(y) \tag{9.8}$$

The annual mean of irradiance $E_{year}(y)$ is given by:

$$E_{year}(m) = H_{year}(y)\big/\left[86,400 N(y)\right] \tag{9.9}$$

In studies on climate change, as the measurements taken on 29 February in leap years are not taken into account, the number of days is the year is always 365. Thus, annual sums and averages can be compared between years.

In meteorology, the World Meteorological Organization recommends proceeding slightly differently when it comes to calculating annual sums and averages. The annual irradiation is the sum of the 12 monthly irradiations for that year and not the sum of the 365 (or 366) daily irradiations:

$$H_{year}(y) = \sum H_{month}(m) \quad \text{for all months } m \text{ in the year } y \tag{9.10}$$

The annual average of the daily irradiation is the average of the 12 monthly averages of daily irradiation and not the average of 365 daily irradiations:

$$H_{annual\_mean\_day}(y) = \left[\sum H_{monthly\_mean\_day}(m)\right]\Big/ 12 \tag{9.11}$$

In doing so, all months have an equal weight in the sum or average regardless of its number of days. Thus, the results will differ depending on how you proceed.

9.4.2 Calculation of climate normals in meteorology

When several years of measurements are available, multi-year averages can be calculated on a monthly or annual basis. Meteorologists have defined the notion of climatological normals or climate normals. For solar radiation, climate normals are annual or monthly averages of daily irradiations averaged over several years. Climate normals have two major purposes: They form a reference against which measured or estimates radiation can be assessed, and they are widely used as an indicator of the solar radiation likely to be experienced at a given location. A technical document of the World Meteorological Organization[6] details how to calculate climate normals for various weather variables and how to use them.

Before going any further, allow me to introduce some notations on the representation of the time intervals according to ISO 8601 standard on dates and times. A time interval, or a period, is defined by a start time and an end time. For dates, ISO 8601 representation of a period is *YYYY-MM-DD/YYYY-MM-DD*. For example, 2011-01-01/2020-12-31 means a time interval starting on 1 January 2011 and ending on 31 December 2020, included. A duration is represented by the letter *P* followed by date and time: *PYYYY-MM-DDThh:mm:ss*. In this example, the duration is 10 years and can be represented by *P0010-00-00* or in short *P0010* or *P10Y*. The period in the previous example may be represented by the start date followed by the duration, such as 2011-01-01/P10Y, or by the duration followed by the end date, such as P10Y/2020-12-31.

The World Meteorological Organization actually defines several normals that differ only in the number of years. The period averages are averages computed over any period of at least 10 years starting on 1 January of a year ending with the digit 1. An example is the period ranging from 2011-01-01 up to 2020-12-31 (2011-01-01/2020-12-31) of duration 10 years (P10Y). Another example is 2001-01-01/2013-12-31 of duration 13 years (P13Y).

Climate normals are period averages computed over a period comprising at least three consecutive 10-year periods, e.g., 30, 40, or 50 years. An example is the period 2001-01-01/2030-12-31 of duration 30 years (P30Y); another one is 1961-01-01/2020-12-31 of duration 60 years (P60Y).

Climatological standard normals are averages computed over specific periods of 30 consecutive years (P30Y). These periods overlap and are updated every decade. Examples are 1961-01-01/1990-12-31, 1971-01-01/2000-12-31, 1981-01-01/2010-12-31, 1991-01-01/2020-12-31, and 2001-01-01/2030-12-31. Thirty years was the number of years having available data when the recommendation on calculation of normal was first made, and this duration P30Y has been kept over time to allow comparisons. Although it is not a formal name given by the World Meteorological Organization, the climatological standard normals for the period 1961-01-01/1990-12-31 are also called climatological standard reference normals. At the time of writing, the current climatological standard normals are for the period 1981-01-01/2010-12-31.

Normals are calculated from monthly values and not individual daily values. Monthly values may be sums (monthly irradiation) or averages (monthly mean of

6 World Meteorological Organization WMO No. 1203, Guidelines on the Calculation of Climate Normals, 2017, 29 p, Geneva, Switzerland.

daily irradiation). A monthly normal for the month m for a period P is the average of all the monthly irradiations or monthly means for this month in this period. For example, the July normal in the current climatological standard normals is the average of the 30 monthly irradiations for July or monthly means in the period 1981-01-01/2010-12-31.

Annual normals are calculated by summing or averaging monthly normals and not from individual annual quantities.

9.4.3 Case of incomplete time series

It often happens that values are missing in the series of measurements of daily irradiation. In this case, what is the significance of the sums and averages calculated with only the measurements available for a month or a year?

As for the sums, it is obvious that they will be less than those calculated with a complete series. Therefore a sum should not be calculated for an incomplete series. However, it is possible to do this if means are available to replace the missing measurements with reliable estimates. This possibility depends on the variability of the radiation, the number of missing measurements, and the distribution of the gaps in the period considered. Let me take two examples to illustrate the point.

In the first example, suppose a time series of daily irradiation $H_{day}(j)$ at a site. The time series lasts one full year starting on 1 January. Suppose 12 days are missing, one for each of the 12 months. In this case, it can often be assumed that the sky conditions, i.e., the atmospheric transmittance, on the missing day for a given month are not much different from those on other days of that month. For each month m, I can calculate the partial average $H_{partial\_mean}(m)$ for the $N_{partial}$ measurements valid for days j in the month:

$$H_{partial\_mean}(m) = \left[\sum H_{day}(j) \right] / N_{partial} \quad \text{for all valid } H_{day}(j) \tag{9.12}$$

Because of my assumption, the partial average $H_{partial\_mean}(m)$ calculated for this month with all the days except the missing one is very close to, if not the same as, $H_{monthly\_mean}(m)$ that would have been calculated in the ideal case of a complete series, with every day valid in the month:

$$H_{monthly\_mean}(m) \approx H_{partial\_mean}(m) \tag{9.13}$$

Then, the monthly sum of the ideal case $H_{month}(m)$ is assumed equal to the estimated average with a missing day, multiplied by the number of days N in that month:

$$H_{month}(m) = N(m) H_{partial\_mean}(m) \tag{9.14}$$

The annual sum is the sum of the 12 monthly sums. In this example, it can be reasonably assumed that the estimate of the annual sum is correct, since each monthly sum is correct.

I can use the same reasoning but with the monthly clearness index $K_{T_{month}}$. Assuming that the sky conditions for the missing day were similar to the average conditions for the month, I can compute the partial average $K_{T_{partial\_mean}}(m)$ and then write:

$$K_{T_{month}}(m) = K_{T_{partial\_mean}}(m) \tag{9.15}$$

$$\text{where} \quad K_{T_{partial\_mean}}(m) = \left[\sum H_{day}(j)\right] \Big/ \left[\sum H_{0day}(j)\right] \quad \text{for all valid } H_{day}(j)$$

$$(9.16)$$

Next, the monthly sum $H_{month}(m)$ is given by the following equation where $H_{0month}(m)$ is the monthly sum of the extraterrestrial irradiation:

$$H_{month}(m) = K_{T_{month}}(m) H_{0month}(m) \tag{9.17}$$

This way of doing, i.e., using the clearness index, is recommended if more than 1 day is missing in a month.

In the second example, I assume that the site is at latitude 65° and the missing 12 days are those at the end of July, from July 20 (day #201 in the year) to July 31 (day #212). This last example may sound like a caricature, but it is not. When daily measurements are missing, it is often a question of technical faults which cannot be solved on the spot and which induce periods of lack of measurements lasting several days. At this site at latitude 65°, July is generally the month with the greatest occurrence of clear skies, i.e., high clearness indices $K_{T_{day}}$. As, moreover, the daily irradiation H_{0day} received by a horizontal plane located at the top of the atmosphere is high in July at this latitude, the irradiations of the missing 12 days have a great influence on the sum of the month of July and consequently on the annual sum. It is impossible to say that the clearness indices of the missing days are similar to, or different from, those of the other 19 days of July. In this case, it is best not to attempt to calculate the sums for both July and the year.

There are several mathematical methods for estimating sums and means in incomplete series. I described two simple ones in the first example above. In the next section, I present techniques for replacing missing measurements with estimates in order to obtain complete series and end up in the ideal case for calculating means and sums.

Each technique has its drawbacks and qualities. What matters most is to take into account the variability of the radiation. You have to be sure that the chosen technique takes into account the specificities of radiation, its different scales in time and space. Generally speaking, the duration or width of a gap or hole must be less than the time or space scale of interest. In the previous second example of the site at latitude 65°, the smallest timescale characteristic of variations in daily irradiation at the top of the atmosphere H_{0day} is about 3 days (see Figure 3.5, Chapter 3). As a consequence, the width of a gap must be less than the minimum between these 3 days and the smallest timescale characteristic of variations in the daily clearness index $K_{T_{day}}$. In the example, the width is 12 days, which is much greater than 3 days; this explains why the sum for July cannot be calculated accurately in the example. For a site located at the equator, the smallest timescale of variations in H_{0day} is around 10 days (see Figure 3.5, Chapter 3), i.e., is close to the width of the 12-day gap. Therefore, the timescale of variations in $K_{T_{day}}$ is decisive in this case.

The World Meteorological Organization recommends calculating the monthly average only if there are at most 10 missing days and no more than 4 consecutive missing days. For my part, I prefer a stricter rule, called the 5/3 rule, adopted by several meteorological services and by the WRDC for its database of daily irradiation. The monthly average is calculated only if there are at most 5 missing days in the month and no more than 3 consecutive missing days.

Reader, you can see that habits and customs vary, including within the same community such as meteorology. I therefore invite you to specify your way of proceeding in your writings and communications.

9.4.4 Overview of the completion of a time series

Completing a series, or replacing missing measurements, is the process of estimating missing values in order to obtain a complete series. This process is called completion or gap-filling. The reasons for carrying out this replacement can be multiple, and the approaches adopted differ according to these reasons. For example, one may want to obtain a denser statistical distribution and thus replace missing values by random draws according to a certain law of probability. Here, my goal is to obtain a series of measurements without gaps in order to describe the variation of the radiation in time or in space.

Any completion or replacement involves assumptions about the variability of the radiation at ground. At a given geographic site, the most influential variables are the extraterrestrial irradiation at normal incidence H_{0N}, the solar zenithal angle θ_S and possibly the angle of incidence on the receiving plane on the ground if the plane is tilted, the constituents of the clear atmosphere and their optical properties, clouds and their optical properties, and the variables describing the reflection of solar radiation by the ground.

Each of these variables has its own time and space scales and its own variability. The variability of the radiation received on the ground is a combination of all the variabilities. It is difficult to know and estimate all these variabilities in order to model that of the radiation at ground to obtain replacement values. One possible approach is to reason on the different scales of each variable in order, on the one hand, to rule out those whose variability is low for the case concerned and, on the other hand, to identify the importance of the others, in order to help choose a mathematical procedure for completion.

Let me take an example to illustrate this point. Consider a series of measurements of irradiation received on a horizontal plane with a regular sampling of 10 s at latitude 40°. Suppose there is a gap of one missing value around solar noon: Two valid 10 s values surround a 10 s gap. Suppose the sky is cloudless and the atmosphere is very quiet and not turbulent. At these timescales of 10–30 s, all the variables show low-intensity variability. This can be checked by comparing the two valid values that should be close to each other taking into account the measurement noise. Then, a replacement value can be proposed which can be equal to one of the two surrounding values or to the average of the two.

Now suppose that the sampling of the series is no longer 10 s, but 1 min. In the absence of further details, it can be assumed that the constituents of the clear atmosphere remain constant over a period of 3 min. Likewise, extraterrestrial radiation H_{0N} and ground reflection vary little on this timescale and can be excluded. The remaining variable is the solar zenithal angle θ_S that varies markedly in 3 min. The completion procedure must take into account the variation of this angle. On a first approximation, one may calculate the average of the two surrounding clearness indices and replace the missing value by the corresponding irradiation. However, if the study is only on the direct component of the radiation received at normal incidence, then the variation of the solar zenithal angle has negligible influence at this scale and the completion procedure can be simplified. For example, a replacement value can be proposed as before, which can be equal to one of the two valid values or to the average of the two.

The most common procedures for completion are persistence and interpolation. Obviously, the larger the width of the gap, the more difficult it is to get an accurate estimate of the missing values. Persistence consists of assuming that the conditions have remained the same and that the missing value is equal to the previous valid value or the next one as appropriate. The mean is a special case of interpolation which can apply when there is only one missing value between two values. In the general case, the interpolation is a weighted sum of the valid values surrounding the gap, the weighting being a function of the time or distance between the middle of the gap and the valid values.

Suppose a series of N values V_i, with $i = 1-N$. Suppose that the three values V_k, V_{k+1}, and V_{k+2} are missing. The missing values can be estimated by:

$$V_k = \left[\sum w_j V_j \right] / \left[\sum w_j \right] \tag{9.18}$$

with $w_j = 1/(j-k)^2$, for example. The farther the time j of the valid measurement V_j from that k of the missing V_k, the weaker the influence of V_j on V_k, and consequently, the smaller the weight w_j.

The completion procedure can be applied to quantities derived from radiation, such as the clearness index, or the clear-sky index. Using these indices is a good way to eliminate variations in extraterrestrial radiation. For example, this makes it possible to overcome variations of E_{0day} with time in the case of a time series of daily irradiation or variations of E_{0day} in space for a spatial series of daily irradiation. One of the variables is thus eliminated, which generally allows less restrictive assumptions to be made in the completion process. If an accurate and reliable clear-sky model is available to estimate E_{clear} or H_{clear}, using the clear-sky index Kc allows the assumptions to be restricted to variabilities in cloud attenuation and reflection by the ground radiation. Approaches using indices can be used at all time and space scales.

Chapter 10

Other means for estimating solar radiation at surface

At a glance

Other ways of estimating radiation have been developed to overcome the difficulties of measuring radiation with ground-based instruments.

Meteorological analyses and reanalyses produce estimates of global, total or spectral radiation, or even its direct and diffuse components. Multispectral imagers on board satellites provide instant views of large geographic areas. Analysis of their images makes it possible to estimate variables related to the attenuation of solar radiation, and consequently, the solar radiation received on the ground. Some methods combine the exploitation of satellite observations and meteorological numerical models.

Spatial interpolation offers the possibility to exploit measurements from stations close to the site of interest or to extend or densify a network of measurement stations in order to produce maps of radiation. There are also empirical methods that convert measured total radiation into radiation in a given spectral range such as UV or visible. Others estimate radiation from other meteorological variables such as sunshine duration, air temperature at surface, cloud cover, or cloudiness.

Chapter 10 is a follow-up of Chapter 9. Both chapters deal with the most common ways to estimate solar radiation received at ground. Chapter 9 presents the direct means for measuring the radiation at ground, i.e., the most common instruments. The present chapter deals with the indirect means.

Earth observation instruments other than dedicated ground-based ones are used to overcome the problem of incomplete coverage in both space and time by stations. They include numerical weather models, meteorological analyses or reanalyses, or readings from instruments on board satellites, including multispectral imagers.

One possible approach is the adoption of so-called direct modeling of radiative transfer. For this, one uses measurements or estimates of all the variables affecting the radiative transfer in the atmosphere, as seen in Chapter 6, and models of the extraterrestrial solar radiation, as seen in Chapter 3. Both are input to a numerical radiative transfer model, in order to estimate the global, total or spectral radiation, or even its direct and diffuse components. This is the case for meteorological analyses or reanalyses, which are the subject of a dedicated section.

Another approach uses Earth observation satellites. Some carry imagers with several spectral bands, called multispectral imagers, and offer synoptic views, i.e., views of large geographic areas on which measurements are acquired at the same time. The processing of measurements, known as inverse modeling, makes it possible to estimate variables linked to the attenuation of solar radiation and therefore to estimate the solar radiation at ground level. Some methods may combine the exploitation of Earth observation data from satellites and the use of meteorological analyses or reanalyses.

The two aforementioned approaches, especially the first, are cumbersome to implement. They often require significant computing resources and also communication resources if measurements or images acquired by satellites are to be used in near real time. Other approaches have been proposed to overcome these implementation difficulties. Some are presented in this chapter.

Spatial interpolation is one of them. It offers the possibility to exploit measurements from stations close to the site of interest or to extend or densify a network of measurement stations in order to produce maps of radiation. Chapter 10 combines the variability seen in Chapter 8, the underlying assumptions, and the uncertainty resulting from spatial interpolation between stations.

There are also methods that I call indirect estimation that involve using measurements or estimates of other variables to obtain the desired information. For this, the indirect measurements must be correlated with the total or spectral radiation sought. These indirect measurements can be sunshine duration, air temperature at surface, cloud cover, or cloudiness in order to assess the total radiation. Or, they can be total irradiations to estimate radiation in a given spectral range such as UV or visible. This chapter does not go into detail as many empirical relationships of this type have been proposed and still are by researchers. I just state the concept and principles and illustrate them.

10.1 Meteorological analyses and reanalyses

Numerical weather models or numerical atmospheric models are an effective way to represent the state of the atmosphere, its motions, its chemistry, and the interactions between the ground, the ocean, and the atmosphere. They are built around the full set of primitive equations that govern the various physical and chemical processes involved. Their inputs are the most accurate and relevant measurements and observations in order to obtain accurate results. These observations are, for example, the air temperature and atmospheric pressure measured at surface or observations of rainfall by weather radars. Vertical profiles of wind speed and air temperature acquired by radiosondes are also part of the observations assimilated by the models among many others. Certain models are capable to assimilate quantities derived from satellite observations such as the optical depths of aerosols or even directly the radiances measured by space sensors. Weather models often provide estimates of the radiation received on the ground, often the total radiation, sometimes also the direct component, as well as extraterrestrial radiation. This is why the outputs of such models are helpful for the knowledge of solar radiation received at ground at any location and at any time.

Some models only deal with the equations of fluid mechanics in the atmosphere and the interactions between the ground, the ocean, and the atmosphere. Others simulate the cycle of chemical species and aerosols, including their advection and deposition. Finally, some simulate all the processes involved. In any case, such models are very complex and are extremely demanding in terms of computer resources; there are few in the world.

Numerical weather models often cover the entire globe or areas of a few thousand km across. The time step is often 1 or 3 h at the time of writing. The spatial step, i.e., the size of a cell, varies between 1 and 100 km. The larger the area covered, the greater the spatial step. Models can be used as tools for forecasting the weather or climate, or analyzing past events.

A weather forecast model provides several outputs, including predicted representations of the state of the atmosphere over the coming hours and days as well as a snapshot at the present time called weather analysis. Analyses are generally archived by their producers. They thus form a time series of maps of the radiation received at ground that could be exploited to study the variability of the radiation in time and space. However, meteorologists regularly update their models to make them more and more accurate. Updating can be changes in model, introduction of new parameterizations, changes in output variables, changes in information support such as a decrease in cell size to improve the spatial representation, and so on. All these changes have an influence on the estimated radiation and make it very difficult to exploit a time series of analyses over several years.

Meteorological reanalyses are one way to overcome these difficulties. A reanalysis is created by reprocessing past weather observations using the same numerical weather model with the same information support. The series of meteorological representations produced is therefore homogeneous in space and time over several decades. Studies of trends and variations in time and space are thus made possible. At the time of writing, the most known reanalyses are likely the Modern-Era Retrospective analysis for Research and Applications (MERRA) and its later versions produced by the National Aeronautics and Space Administration of the United States, the ERA-5 and its later versions produced by the European Centre for Medium-Range Weather Forecasts in Europe, or the Japanese 55-year Reanalysis (JRA-55) produced by the Japan Meteorological Agency.

For both analyses and reanalyses, it should be noted that the variables describing the radiation are diagnostic. There is no assimilation of measurements of radiation. The variables derive from a radiative transfer model. They are therefore subject to errors, due in particular to simplifications of the model for estimating cloud properties. Comparisons between radiation measurements made by pyranometers on the ground or at the ocean surface and reanalysis results show that reanalyses tend to underestimate or overestimate the presence of clouds and their optical depth. Clear-sky conditions are estimated to be cloudy and vice versa. As a result, intraday and day-to-day variations in solar radiation over time are not well reproduced by reanalyses. Overestimates and underestimates may eventually compensate for each other depending on climate and geographic areas, in which case the annual averages may be close to those measured by pyranometers.

Numerical weather models are regularly improved, from the point of view of both the modeling of physical processes and their ability to assimilate more and more Earth observation data – in the sense of the Group on Earth Observations – coming from more and more instruments, with increasingly fine spatial and temporal resolutions. Estimates of solar radiation received on the ground will become more and more reliable and accurate in the years to come, and I believe that reanalyses will be used more and more in the field of solar radiation. At the time of writing, it is best to use estimates derived from processing images acquired by Earth observation satellites, if possible.

Reanalyses have larger spatial and temporal coverages than satellite estimates. It is tempting to merge or fuse reanalyses and satellite-derived estimates to get the synergy of the qualities of each data set: the spatial and temporal coverages of the reanalyses and the better accuracy of the satellite-derived estimates. The concept of the fusion process is to correct the reanalysis of interest with the coincident satellite-derived estimates and then extend that correction to the entire reanalysis to obtain a corrected reanalysis. The data input to a fusion process must be aligned;[1] i.e., the information supports of the two data sets must be the same in time and space. Alignment is usually done by modifying the information supports of the satellite-derived estimates to coincide with those of the reanalysis. Once the alignment is obtained, the two data sets can be compared and a calibration function can be established, which brings the reanalysis irradiances closer to those estimated by satellite for the coincident period. The calibration function can be, for example, an affine function or an abacus created by an adjustment of the quantiles of the statistical distributions of irradiances or of the clearness indices. Finally, the calibration function is applied to all the reanalysis irradiances to obtain a corrected reanalysis, assuming that this function is valid for the entire period and for the entire globe.

10.2 Images acquired by Earth observing satellites

As already mentioned in the previous section, data acquired by the various space-borne instruments are widely used in numerical meteorological models. These models produce estimates of many meteorological and atmospheric chemistry variables, including solar radiation incident on the ground. The latter can also be estimated directly from images acquired by satellite. Spaceborne imagers provide so-called synoptic views, i.e., views of large geographic areas for which pixels can be deemed to have been observed at approximately the same instant. In other words, the image is a snapshot of the large geographic area. These imagers therefore make it possible to estimate the solar radiation received on the ground and its spatial distribution at a given instant. In this section, I deal with multispectral imagery, then with the orbits of the satellites carrying the imagers. Finally, I present the general principles of the main methods of estimating radiation from satellite imagery.

---

1 See e.g. Wald L., 2002. *Data Fusion. Definitions and Architectures – Fusion of Images of Different Spatial Resolutions*, Presses de l'Ecole, Ecole des Mines de Paris, Paris, France.

10.2.1 Multispectral imagery

Multispectral imagers on board satellites observe in several spectral bands. They collect radiance from various bodies on the Earth, the atmosphere, and its constituents, including clouds.

As seen in Chapter 3, any body emits radiation provided that the temperature of its surface is above 0 K (approximately −273 °C). The radiation is emitted in a spectral interval which is a function of the surface temperature. The wavelength of maximum emission intensity and the intensity are also a function of this temperature. This is the case of the sun whose surface temperature is very high, around 5780 K (around 5500 °C), and whose emission spectral interval is approximately [200, 4000] nm. The lower the surface temperature, the greater the wavelength of maximum emission intensity. The various constituents of the Earth, whether they are solid like the ground, liquid like the ocean, or gaseous like the atmosphere in general, have much lower surface temperatures than the sun, and the emission maxima take place for wavelengths of about 10 μm. The emission spectral interval is approximately [3, 80] μm. Imagers capture the radiances emitted by these objects in the thermal infrared bands, between 8 and 12 μm. As a first approximation, the measured radiances are a function of the surface temperature of the bodies. The greater the temperature, the greater the measured radiance. In reality, one must take into account the emissivity, which depends on the observed object and the observation geometry.

In wavelength bands less than 1 μm, imagers measure the radiance due to the reflection of solar radiation by terrestrial and atmospheric objects. At wavelengths close to 3 μm, the measurement is a mixture of the emitted and reflected radiances. Reflected radiance was discussed in Chapter 5 mainly, with in particular the notion of albedo of the atmosphere and bidirectional reflectance distribution function. The more reflective the body, the greater the radiance. For example, except specular reflection conditions, radiances reflected by the water surface in deep water are significantly less than those reflected by deserts, snow-covered surfaces, or clouds.

An object emits and reflects different radiances at different wavelengths, and this spectral distribution is different from that of another object. Observation of an object by a multispectral instrument provides information on the nature of that object and its properties. Different types of soil, vegetation, or clouds, can thus be characterized.

One of the major difficulties in the use of multispectral images acquired by satellites is that the observed signal results from a complex combination of interactions between the solar radiation incident at the top of the atmosphere, the atmosphere, and the ground in the downward direction, i.e., toward the ground as seen in Chapter 6, and between the radiation emitted or reflected upward toward the satellite through the atmosphere. Unlike so-called direct modeling of the radiative transfer used in numerical meteorological models, which deals only with the downward path toward the ground, the two paths must be taken into account here: upward and downward. This modeling that aims at obtaining information on the interactions and underlying processes from the observed radiances is called inverse modeling. An accurate estimate of the radiation received on the ground requests the ability to estimate the interactions between the radiation and the atmosphere during the upward path, and for this, to have sufficient information on the thermo-optical properties of the objects, on the constituents of the atmosphere, including clouds and vertical distributions of atmospheric properties.

10.2.2 Orbits of satellites

Satellites are in orbit around the Earth. There are generally two types of orbit in satellites used operationally in meteorology and more generally in Earth observation: the near-polar orbit and the geostationary orbit. Both are quasi-circular orbits.

Satellites in near-polar orbits are at altitudes of about 800 km. The duration of an orbit is approximately 100 min or 1 h and 40 min. These orbits are such that they pass near the poles; the trajectory of the satellites is essentially from south to north, called ascending direction, then from north to south, known as descending direction, after having crossed the North Pole. Orbits are often sun-synchronous; that is, the satellite crosses the equator at about the same time in mean solar time (MST) for the pixel below the satellite. The times chosen are generally mid-morning, between 09:00 and 10:00, or afternoon, between 15:00 and 16:00 MST. Since the Earth rotates during the revolution of the satellite, the satellite does not pass over the same geographic point on the equator from one orbit to another. There is a shift of about 11° in longitude to the west. In other words, on the next revolution, the satellite passes over a point about 1200 km further west than the previous one.

At one time, the imager observes a portion of the surface of the Earth. The swath is the width of this portion in the direction orthogonal to the movement of the satellite. At the equator, multispectral imagers for meteorology have swaths of about 3000 km centered on the satellite track on the ground, or equivalently 1500 km on either side of the satellite track. Figure 10.1 is a schematic view of the ground track of a near-polar

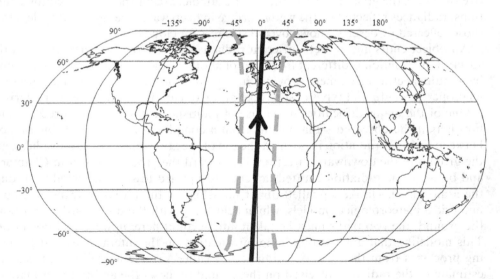

Figure 10.1 Example of the ground track (solid thick line) of a near-polar orbiting satellite and swath limits (dotted lines) of a spaceborne imager having a swath width of 3000 km at equator.

orbiting satellite and of the swath limits of a spaceborne imager having a swath width of 3000 km at equator. In this graph, the satellite is ascending. At the equator, a degree in longitude corresponds to approximately 110 km; i.e., a swath of 3000 km corresponds to approximately 27° in longitude. The swath of the imager increases toward the poles because of the Earth curvature. In Figure 10.1, it is of the order of 90° at poles.

Given a geographic point, how often does the imager on board the satellite see it while the sun is above the horizon? Take the case of a point at the equator. Suppose the satellite goes into ascending orbit just above this point at 09:30 MST. The point is seen in the middle of the swath. Given the large swath, it has already been seen for the first time, on the western part of the swath, 100 min earlier, i.e., at around 07:50 MST. At that time, the satellite has crossed the equator about 1100 km further east. The point will be seen a third time, on the eastern side of the swath, at approximately 11:10 MST. Then, with the orbit shifting further, it will not be seen until the next day. It will actually be seen overnight three times around 19:50, 21:30, and 23:210 MST, but the sun does not illuminate the scene at those times. Three observations per day are therefore available for this point. On the other hand, a geographic point located at high latitudes is seen much more frequently. Indeed, at these latitudes, a shift of 11° in longitude to the west no longer represents 1100 km but much less, for example, 550 km, at latitude 60°. At this latitude, the point is seen approximately five times per day.

The temporal sampling of imagers on board near-polar orbiting satellites is irregular and depends on the latitude. An equatorial area will be seen less often than an area at high latitudes. The presence of several satellites (2 or 3), with similar orbits but out of phase in time and equipped with similar multispectral imagers, significantly increases the frequency of observation of an area.

The geostationary orbit is such that a satellite on it has the same speed of rotation as that of the Earth on itself. In other words, the satellite is always above the same geographic point. This choice of orbit requires that this geographic point be located on the equator. It only takes five geostationary satellites to cover the entire globe, with the exception of the polar areas. Figure 10.2 is an example of images acquired by an imager on board a geostationary satellite, here the European Meteosat satellite located at longitude 0°.

The Earth appears as a disk. The part outside the Earth rim is black because the radiance from the space is zero. The radiance increases from black to white. Africa is the white mass in the middle of the image with high measured radiance. To the north stands out Europe, separated from Africa by a dark area, which is the Mediterranean Sea. The Arabian Peninsula and Asia Minor may be seen to the east. The Atlantic Ocean is the great dark mass west of Africa. The eastern part of South America is clearly visible. Clouds appear in different places, showing different shades from dark gray to white.

Reader, as an astute observer, you have noticed that the details are less visible toward the outside of the image than in its center. The size of the pixel increases very quickly when the angle of view increases, i.e., as it moves away from the center of the image, and this decreases the perception of details. On the edges of the terrestrial disk, the line of sight is tangent to the Earth and details cannot be distinguished any longer. These orbits do not make it possible to obtain images of the polar zones, typically beyond latitudes −65° and +65°.

Figure 10.2 Example of images acquired by meteorological geostationary satellites. Here, a composite image made from multispectral images obtained by the Meteosat satellite on 2019-06-21 at 12:12 UTC is shown. Copyright (2019) EUMETSAT.

Temporal sampling is regular for imagers on board geostationary satellites. It is around 10 min at the time of writing. Since the viewing geometry is almost always the same, the pixel size at a given location does not vary from shot to shot as in the case of a polar satellite. In a time series of images, the same geographic area is made up of the same pixels, each retaining its size in the series. The uncertainty on the geographic position of each pixel must be taken into account, which is of the order of 1/3 of a pixel.

10.2.3 Overview of the methods for estimating the solar radiation

Many methods have been published for estimating the solar radiation received at ground from images acquired by spaceborne multispectral imagers. I do not intend to make a review. Some authors have already done it very well; others will also do it very well. Among the scientific publications, I can identify three different general concepts, which apply to both types of orbits. The problem is to estimate the solar radiation having been received on the ground during the downward path by knowing the radiance received at the satellite level after an upward path. The two really known quantities are the radiance measured by the imager and the extraterrestrial radiation.

In the first two concepts, the downward and upward paths are assumed to have the same properties with respect to radiative transfer, which is more or less true as long as the solar zenithal and viewing angles are not very large. The extinction of solar radiation by the atmosphere can then be estimated. These two concepts often involve the decoupling of the effects of clouds from those of the clear atmosphere, seen in Section 6.7 in Chapter 6. Let E_0 be the solar irradiance at the top of the atmosphere, E the irradiance received on the ground, E_{clear} that received on the ground in cloudless conditions, and Kc the clear-sky index. Let P_{clear} be the set of variables describing the constituents of the cloudless atmosphere and P_{clear_type} a standard or average vector of variables describing the clear atmosphere, chosen arbitrarily from the P_{clear} set. Finally, let P_{cloud} be the set of variables describing clouds. Then, the irradiance E is given by:

$$E \approx E_{clear}\left(E_0, z, \theta_S, \rho, P_{clear}\right) Kc\left(z, \theta_S, \rho, P_{clear\_type}, P_{cloud}\right) \tag{10.1}$$

where z is the elevation of the ground above the mean sea level, θ_S is the solar zenithal angle, and ρ is the bidirectional reflectance distribution function.

This approximation is valid for total radiation as well as for narrower spectral intervals. Thanks to it, the radiation calculations can be decoupled and simplified: on the one hand, the calculation of $E_{clear,}$ and on the other hand, that of Kc. One can call upon a more or less complex, but already existing clear-sky model to calculate E_{clear} like those used in the previous chapters. The inputs to these models usually come from sources other than the multispectral images under concern. Some clear-sky models have only the solar zenithal angle as input. Others use mean values for aerosols or Linke turbidity factor. Still others use outputs of numerical weather models such as the Copernicus Atmosphere Monitoring Service of the European Union. Once E_{clear} is calculated by the clear-sky model, the problem is limited to estimating the clear-sky index Kc from images acquired by satellite. The first two concepts tackle this calculation of Kc.

The first concept is based on the cloud index, often noted n. Note that in this concept, the possibilities of multispectral imaging are not fully exploited. The first step is to construct one or more time series of satellite observations at each pixel from which the minima and maxima can be extracted. It is recommended to build at each pixel several time series of radiances L_{sat} measured by the imager. Using radiances instead of digital counts output by the sensor overcomes the problems induced by changes in calibration or sensors. Each series is made of radiances acquired in the same relative sun–Earth–satellite geometry in order to overcome the effects of changes in reflection due to changes in the geometry of illumination and observation. Series are limited to one phenological season of plants to limit the influence of changes in plants on reflection. They can possibly cover the same period over several years. Since, except in special cases, the radiance of soils and water bodies is less than that of clouds, the minimum radiance L_{min} of the series corresponds to an observation of the ground in cloudless conditions and the maximum L_{max} corresponds to the observation of a very reflective cloud of high optical depth. In other words, given the radiance L_{sat} observed at a given pixel and at a given time, the atmosphere is likely cloudless if L_{sat}

is close to L_{min}. Conversely, the presence of an optically very thick cloud is probable when L_{sat} is close to L_{max}. These quantities, L_{min} and $L_{max,}$ are established prior to the routine use of the images. Reader, you have certainly noticed in Figure 10.2 that deserts have radiances close to those of clouds, or even greater. Techniques exist to deal with these particular cases. I do not detail them here.

The cloud index n is defined as the difference between the radiance L_{sat} observed at a given pixel and at a given time and L_{min}, normalized by the difference between the maximum L_{max} and the minimum L_{min}:

$$n = \left(L_{sat} - L_{min} \right) / \left(L_{max} - L_{min} \right) \tag{10.2}$$

When L_{sat} is close to L_{min}, it means that the sky conditions are close to the cloudless conditions and n is close to 0. On the contrary, when L_{sat} is close to L_{max}, optically thick clouds are present and n is close to 1. The behavior of n is opposite to that of the clear-sky index Kc, which is 1 in the case of cloudless skies and is closer to 0 when the attenuation of radiation by clouds is strong. As a first approximation, $n \approx 1 - Kc$. The methods using the cloud index are fairly simple to implement and provide very good results. Total or spectral radiation can be calculated directly in this way, but not the direct and diffuse components which must be calculated indirectly from the global radiation as seen later.

The second concept implements a radiative transfer model more explicitly than in the case of the cloud index. Knowing that the major influence on Kc is due to clouds, multispectral imagery is exploited to extract information on the properties of clouds, such as their optical depth, the pixel coverage rate, their phase (liquid or ice), their altitude, liquid water content, or the effective diameter of water droplets and ice crystals. This information is then entered into the radiative transfer model that calculates Kc. The model is often represented by pre-calculated abaci or charts, also known as look-up tables, in order to reduce the computer resources required for a calculation. The global, direct and diffuse clear-sky indices are obtained in a single calculation, thus yielding the global radiation and also the direct and diffuse components if the clear-sky model provides these components.

The third concept is to create a database of multispectral radiances that can be observed by the considered multispectral imager in all possible situations. This database is created by running a numerical radiative transfer model by varying all the variables having an influence on the solar radiation received on the ground and reflected or emitted in order to calculate the multispectral radiances. The database includes also the radiation received at ground together with its direct and diffuse components. When a multispectral observation is available, it is compared to those stored in the database. The closest is found and one reads the associated radiation received at ground and its components.

All three concepts can be used to calculate total radiation or radiation in any spectral range such as ultraviolet radiation, or daylight or photosynthetically active radiation.

Of course, quite similar to what has been discussed for reanalyses, the methods can also be improved by a correction or calibration function estimated from measurements of equivalent radiation made at ground stations. The estimation of the calibration function can be done daily or at other time intervals, or once and for all. The difference between the measurements and the satellite-derived estimates is modeled to obtain a correction function covering the geographic space to correct all pixels and the entire time period. The solar zenithal angle and the clearness index are explanatory variables often used in this modeling of the difference.

10.3 Estimate using nearby ground stations

The use of reanalyses or satellite imagery can be quite cumbersome to implement when expertise with these products and computer resources are limited. Other solutions are possible, in particular those aiming to exploit the measurements of ground stations close to the site of interest. The time series measured at these stations are combined into a single time series, which then serves as a substitute for that which could have been measured at the site had it been equipped with the same instruments. Many published works have explored the possibilities offered by spatial interpolation and extrapolation of the measurements acquired by nearby ground stations.

There must be a relationship between the solar radiation received on the ground at the site of interest and that received at neighboring stations. How do you know? For a link to exist, the stations and the site must experience the same climates from the point of view of the radiative transfer; i.e., the atmosphere is the same above these geographic points and they see approximately the same clouds, the same aerosols, and other constituents of the atmosphere. In the event of marked relief, the stations and the site must be subjected to the same shade. The surrounding soil types must also be similar in terms of reflection. I presented in Chapter 5, the example of Thartar Lake, in Iraq, which is surrounded by a large desert area, erg type. I showed by a quick calculation, that the relative difference in irradiance observed of about 5 % between the center of the lake and a site located about 10 km inland, can be explained by the differences in ground reflection. The nature of the ground should therefore not be overlooked.

Suppose there is a link. How do you know if a station can be qualified as close to the site in question? This is where the variability seen in Chapter 8 comes in. For a given distance d separating the station from the site, the greater the spatial variability of the radiation, the weaker the relationship between the irradiation received at the site and that measured at the station. The sampling of measurements at the station, or the integration time, must be taken into account in order to determine the observable timescales. Since timescales are related to spatial scales, the latter can be approximated. The comparison between these scales with the distances between the site and the stations, defines the nearby stations. In Chapter 8, I presented the case of a very dense network, with sites spaced 2 km apart. I have found that the relative error made by assuming that the hourly irradiation received at the site of interest is identical to that received at a station 10 km away, can reach 28 %. If the distance is 100 km, the observed relative error is 40 %. The more remote the station, the greater the relative error made in estimating that the hourly irradiation received at the site is identical to that of the station. The errors given here are only realistic examples and should not be considered general. They depend on the variability that strongly depends on the geographic region of interest. The variability also depends on the timescale considered. Therefore, relative errors also depend on it. For the same distance d, the relative error is much less for daily irradiations than for hourly irradiations. The errors decrease further for monthly or annual irradiations. Correlatively, this means that one can choose stations more distant from the site for monthly irradiations than for hourly irradiations with the same relative error. In general, spatial interpolation only gives good results if the stations are close enough and numerous enough for the scales of variation of interest to be represented in the measurements.

Spatial interpolation is a mathematical process by which the radiation is estimated at a point and a time using a network of stations. Extrapolation is an extension of interpolation for which the considered site is not inside the zone defined by the stations, but outside. The mathematical process is often the same, and here I am deliberately confusing interpolation and extrapolation. A large number of interpolation techniques have been proposed. I just state the concept of interpolation and illustrate the principle of linear interpolation as an example (Figure 10.3). Given three measuring stations MS1, MS2, and MS3 for which the irradiations H1(t), H2(t), and H3(t) are known, the objective is to estimate the irradiation H(t) at location P surrounded by the three stations.

The principle of linear interpolation is to take into account the distance d between the point of interest and each of the stations to weight all the measurements and obtain an estimated series of irradiation. The smaller the distance d, the stronger the influence of the station. In the general case where N measuring stations MS_i for which the irradiation H_i is known are surrounding P at the distance d_i from P, the irradiation H is given by:

$$H = \left[\sum w_i H_i\right] / \left[\sum w_i\right], \quad \text{for } i = 1 \text{ to } N \tag{10.3}$$

In this equation, w_i are the weights of the irradiations H_i. Weights are function of the distances d_i, so that the greater the distance d_i, the smaller the weight w_i. The scientific literature is full of relationships between d_i and w_i. The so-called gravity relation, $w_i = 1/d_i^2$, is often used.

Likewise, the scientific literature offers various expressions for the distance between point P and station S_i. One of the most common is geodetic distance. Let (Φ_p, λ_P) be the pair (latitude, longitude) of P, and let $(\Phi_{MSi}, \lambda_{MSi})$ be the pair of MS_i. The mean

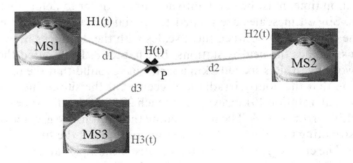

Figure 10.3 Illustration of the spatial interpolation to estimate radiation H(t) at location P. Measurements H1(t), H2(t), and H3(t) are acquired at measuring stations MS1, MS2, and MS3, respectively. d1, d2, and d3 are the distances between P and the stations.

latitude Φ_{mean} is equal to $(\Phi_p + \Phi_{MSi})/2$. Let R_{Earth} be the mean radius of the Earth, i.e., 6368 km. The geodetic distance d_{geo} is given by:

$$d_{geo}^2 = R_{Earth}^2 \left[\left(\Phi_p - \Phi_{MSi} \right)^2 + \left(\left(\lambda_p - \lambda_{MSi} \right) \cos \left(\Phi_{mean} \right) \right)^2 \right] \tag{10.4}$$

Other expressions make it possible to take into account not only the geodetic distance but also the difference in elevation Δz between P and the station MS_i, such as for example, this distance:[2]

$$d^2 = f_{NS}^2 \left[d_{geo}^2 + f_{elevation}^2 \Delta z^2 \right] \tag{10.5}$$

with $f_{NS}^2 = 1 + 0.3 \left| \Phi_p - \Phi_{MSi} \right| \left[1 + \left(\sin \left(\Phi_p \right) + \sin \left(\Phi_{MSi} \right) \right) / 2 \right]$ and $f_{elevation} = 500$; latitudes are expressed in sexagesimal degrees and Δz in km.

Interpolation techniques can be used for total radiation or radiation in narrower spectral intervals. If the stations are a little far apart, in the case of intraday irradiations, it must be taken into account that the solar zenithal angle is not the same at P and at each station MS_i at a given time. To avoid this problem, I recommend to interpolate the clearness index K_T, then to multiply the result by the extraterrestrial irradiation H_0 at the site to obtain the interpolated irradiation H. This amounts to saying that the attenuation of the atmosphere is the same for the same legal time at P and at the stations, which is less restrictive than saying that the irradiation is the same. If a good clear-sky model is available, it is best to interpolate the clear-sky index K_c. If measurements to interpolate are daily, weekly, monthly, or annual values, the variation of the solar zenithal angle does not matter and one can as well use the irradiation (or the irradiance) that the clearness index.

10.4 Usual approach to estimate components from the only knowledge of global radiation

Very often, the radiation measurements available are global irradiations on a horizontal plane, possibly transformed into global irradiances. Measurements of the direct and diffuse components are much less frequently available, either for total radiation or for radiation in a narrower spectral interval. This is also true for the radiation estimates provided by meteorological analyses or reanalyses. However, the components B and D on the horizontal plane may be needed, for example, to estimate irradiance within a foliage or in a greenhouse or for concentrated solar systems for the production of heat, or to know the components B, D, and R on an inclined surface: a hill, a greenhouse, a window, and a photovoltaic panel, for example.

---

2 Lefèvre M., Remund J., Albuisson M., Wald L., Apr 2002. Study of effective distances for interpolation schemes in meteorology. European Geophysical Society, 27th General Assembly, Nice, France. European Geophysical Society, EGS02 (2), pp. A-03429. [Several distances are assessed for estimating meteorological variables at surface, including solar radiation, sunshine duration, air temperature, atmospheric pressure, relative humidity and amount of rainfall, available at https://hal-mines-paristech.archives-ouvertes.fr/hal-00465570/document.]

Consequently, a great activity of researchers, students, engineers, and other practitioners consists in estimating the components from the sole knowledge of the global radiation. Numerous scientific and technical publications have been devoted to this subject and still are. It is not my goal to detail one or more of these models. There are so many! Some are more appropriate than others in a given region, for a given spectral interval, or for a given integration time (e.g., minute, hour, day, month). Others use available measurements or other meteorological variables. However, I describe below the usual approach to estimate the radiation on an inclined plane, which is generally in two steps.

In the first step, a model provides the direct $B(0, 0)$ and diffuse $D(0, 0)$ components on the horizontal plane from the global irradiation $G(0, 0)$. Such a model is often called decomposition model. Inputs may include the clearness index, cloud cover, sunshine duration, or other quantity resulting from a weather type classification, and the solar zenithal angle in case of intraday radiation.

In the second step, each component is treated separately to estimate the direct $B(\beta, \alpha)$ and diffuse $D(\beta, \alpha)$ components on the inclined plane. At a given time t, the direct component $B(\beta, \alpha, t)$ is linked to the direct component $B(0, 0, t)$ received by a horizontal plane (Figure 6.1) by:

$$B(\beta,\alpha,t) = B(0,0,t)\cos\big(\theta(t)\big)\big/\cos\big(\theta_S(t)\big) \tag{10.6}$$

During the integration time $[t_1, t_2]$ corresponding to the measurement $G(0, 0)$, the angles θ and θ_S vary. It is therefore necessary to use effective angles θ^{eff} and θ_S^{eff} as seen in Chapter 3 but applied to the case of the inclined plane. Assuming that there is no shading on the plane during the integration time, the results obtained in a previous chapter about the irradiation $H_0(\beta, \alpha)$ received on a plane inclined at the top of the atmosphere during a time interval (equation 3.24) can be used. It comes:

$$B(\beta,\alpha) = B(0,0)\big[H_0(\beta,\alpha)\big/H_0(0,0)\big] \tag{10.7}$$

The estimation of the diffuse component $D(\beta, \alpha)$ is much more complex and uncertain. The main cause of the difficulties is the lack of knowledge of the angular distribution of radiances on the sky vault for the case under concern. This distribution is rather isotropic when the sky is uniformly covered with clouds optically thick enough to completely mask the sun (absence of the direct component). It is generally anisotropic with a marked circumsolar effect if the sky is not too cloudy. The presence of clouds scattered or grouped together in certain portions of the sky accentuates this anisotropy. Models of angular distribution of radiances are regularly published. They depend on the state of the atmosphere, the position of the sun, and the spectral intervals involved. They can be used for the estimation of the diffuse component received by the inclined plane.

A very simple model consists in assuming that the angular distribution of radiances on the sky vault is uniform. The fraction $r_{sky}(\beta)$ of sky seen by the inclined plane (Figure 6.1) is:

$$r_{sky}(\beta) = \big(1 + \cos(\beta)\big)\big/2 \tag{10.8}$$

If $D(0, 0)$ denotes the diffuse component received by the horizontal plane, a rough approximation of the diffuse component $D(\beta, \alpha)$ is given by:

$$D(\beta,\alpha) = D(0,0) r_{sky}(\beta) \tag{10.9}$$

The irradiation $H(\beta, \alpha)$ is equal to the sum of $B(\beta, \alpha)$ and $D(\beta, \alpha)$, to which the reflected component $R(\beta, \alpha)$ is added as discussed in Section 6.1.1 in Chapter 6.

10.5 Methods using other meteorological variables

Many methods have been developed and published, which use measurements or estimates of other meteorological variables to estimate radiation. This is what I call an indirect estimate because the other meteorological variables are used as substitutes to the solar radiation. The meteorological variables are measured at the site of interest or at other stations. In the latter case, it is possible to perform a spatial interpolation of each meteorological variable; then, the radiation is estimated by the chosen indirect method. Or the radiation can be estimated at each station by the selected indirect method and then a spatial interpolation of the radiation at the site of interest is performed.

The estimated radiation can be total radiation or radiation in a narrower spectral range: For example, many relationships are published concerning the radiation in the ultraviolet or the photosynthetically active radiation. Indirect measurements can be sunshine durations or surface air temperatures or cloudiness or others. Relationships are generally empirical and have limited geographic areas of application. I present here the general features of the most usual methods.

Since these methods relate to data fusion, it is first of all necessary to verify that the chosen meteorological variables and the total or spectral irradiance sought are correlated. For example, the difference between daily maximum and minimum of surface air temperature is well correlated with daily irradiation. As a counterexample, I may cite the average daily surface wind speed that is generally poorly correlated with daily irradiation; there is therefore no point in looking for a relationship. Since the variability of a meteorological variable is specific to it, this correlation can vary depending on the timescale considered. Great attention must be paid to the latter. For example, there is little correlation between the hourly sunshine duration and the hourly irradiation, but the correlation is strong between monthly values. The variables have also to be aligned, i.e., share the same time support of information.

Empirical methods include parameters that often depend on geographic location and the duration of integration. Different sets of parameters between, for example, hourly irradiation and daily irradiation, are usually proposed for the same mathematical formulation.

10.5.1 Ångström's relationship between irradiation and sunshine duration

One of the best known empirical relationships is the Ångström relationship, also known as the Ångström–Prescott relationship. It links the monthly mean of the daily irradiation to the monthly mean of the daily sunshine duration. First of all, remember that the sunshine duration S is the duration during which the direct component of the irradiance received at normal incidence on a plane on the ground is greater than 120 W m^{-2}. It is expressed in h and was defined in Chapter 6. The sunshine duration is generally measured over the day with a device called a sunshine recorder. Sunshine recorders are automatic devices and more widespread than pyranometers in many countries.

The ratio of the daily sunshine duration S to the astronomical daytime S_0 is called the relative sunshine duration. As a first approximation, the smaller the relative sunshine duration, the greater the extinction by the atmosphere, and the smaller the clearness index K_T. Ångström's relationship was established by taking advantage of this correlation. Let H be the irradiation received on the ground on a horizontal plane and H_0 that at the top of the atmosphere. This relation links the H/H_0 ratio to the relative sunshine duration:

$$K_T = H/H_0 = a + b\left(S/S_0\right) \tag{10.10}$$

This relationship is very practical and often used. However, it has limitations. In particular, as S depends only on the direct component B_N of the radiation, while H is the sum of the direct component B and diffuse D, S can be insensitive to variations of H induced by variations of D only. In other words, the correlation between variations in S and H is not equal to 1. This means that this relation does not make it possible to represent the variations in radiation at all scales. This is why it is recommended to use this relationship with caution and preferably to estimate monthly or annual irradiations, or monthly or annual averages.

The parameters a and b are computed at any measuring station for which time series of the two variables H and S are available. They are assumed to be constant in a geographic area around the station. Therefore, any time series of S acquired at any site in this area can be converted into a time series of irradiation using this couple of parameters. In the case where the relationship has been established at several stations and where there are therefore several couples (a, b), the radiation can be estimated at a distant point by spatial interpolation or extrapolation. Please note that you should not interpolate the parameters a and b. You have to calculate H at each station at all times using the corresponding couple (a, b) and then perform the spatial interpolation of H at the site of interest. If $b = 0$, then the operation is simpler as you may interpolate a to the site of interest and then apply the relationship.

The parameters a and b can be evaluated approximately at any geographic point where S is known without needing to know H. Indeed, when the duration of sunshine S is zero due to the cloudiness or the turbidity of the cloudless atmosphere as in the case of a dust storm, then the relationship is reduced to $H/H_0 = a$. The parameter a is therefore the minimum of the H/H_0 ratio and is equivalent to the minimum daily clearness index K_{Tmin}. K_{Tmin} can be set equal to 0.03 everywhere as a first approximation, according to various published estimates. In addition, if the sky is very clear, the durations S and S_0 are equal and the sum $(a+b)$ corresponds to the maximum daily clearness index K_{Tmax} observed at the location considered for the month considered. The relationship can then be written:

$$H/H_0 = K_{Tmin} + \left(K_{Tmax} - K_{Tmin}\right)\left(S/S_0\right) \tag{10.11}$$

KT_{max} varies according to the place, the month, and the constituents of the clear atmosphere. There is no single value. The daily KT_{max} in a given month can be estimated using a clear-sky model.

Many scientific publications have proposed expressions of a and b as a function of latitude, longitude, and elevation above the mean sea level. Other modifications have been made to the original formula by including other variables, such as for example, cloudiness or relative humidity at the surface.

10.5.2 Relationships between the total radiation and that in a given spectral range

Scientific publications give numerous empirical relationships between total irradiance and irradiance in a narrower spectral range such as irradiance in the ultraviolet or photosynthetically active radiation. There are not many stations equipped with pyranometers measuring total radiation in many places. However, they are more numerous than those measuring ultraviolet radiation or photosynthetically active radiation. This is why these relationships estimating radiation in a spectral range from total radiation measurements have been developed. For example, a relation of the following type is often used in agro-meteorology:

$$E_{PAR} = aE_{pyranometer} \tag{10.12}$$

where E_{PAR} is the photosynthetically active irradiance and $E_{pyranometer}$ the irradiance measured by a pyranometer. a is a factor that is typically between 0.45 and 0.52 depending on the authors. These values of a are approximately those reported in Table 7.2 (Chapter 7) in clear-sky conditions (0.44 and 0.45) and in Table 7.4 in cloudy conditions (0.46 and 0.51).

Relationships are not necessarily linear, like the relationship above, or affine functions, like Ångström's relationship. The parameters of a relation are determined empirically at stations where all the variables and radiations entering into the relation are known. The geographic area of applicability is given by these stations, their climate, and their geographic distribution. These relations are approximations, and the parameters generally depend on the sky conditions, which can be represented by the relative sunshine duration, the clearness index KT, or the cloudiness, and on the solar zenithal angle for intraday timescales. For example, there are many relationships of affine type linking the hourly averages of ultraviolet irradiance E_{UV} and $E_{pyranometer}$:

$$E_{UV} = aE_{pyranometer} + b \tag{10.13}$$

where the parameters a and b are tabulated and depend on the solar zenithal angle.

Some relationships aspire to geographic universality by taking into account one or more variables relating to sky conditions. For example, the European Solar Radiation Atlas (ESRA) gives a relationship for estimating the spectral irradiance $E(\lambda)$ from $E_{pyranometer}$ as a function of the relative sunshine duration:

$$E(\lambda) = \left[1.163 \, 10^{-5} (\lambda - 300)\right] f(\lambda) E_{pyranometer}, \quad \text{if } 310 \, \text{nm} \leq \lambda \leq 465 \, \text{nm} \tag{10.14}$$

$$E(\lambda) = \left[3.1515 \times 10^{-3} - 2.651 \times 10^{-6} \lambda\right] f(\lambda) E_{pyranometer}, \quad \text{if } 465\,nm < \lambda \leq 1000\,nm$$

with

$$f(\lambda) = \left(1 - f_{clear}(\lambda)\right)\left(S/S_0\right) + \left(1 - f_{cloud}(\lambda)\right)\left[1 - \left(S/S_0\right)\right] \qquad (10.15)$$

where the correction factors $f_{clear}(\lambda)$ and $f_{cloud}(\lambda)$ are given in Table 10.1. Estimation errors (root mean square error) can reach 20 % in relative value for hourly average of irradiance.

This ESRA relationship may be used to compute the irradiance over a given spectral range between 310 and 1000 nm. Let this range be $[\lambda_1, \lambda_2]$. The relationship is applied to each wavelength in this range every 10 nm, and the resulting irradiances are multiplied by 10 nm, which is the elementary width in this approach, and finally summed up to yield the spectrally integrated irradiance $E[\lambda_1, \lambda_2]$.

The ESRA relationship can be linked to the Ångström relationship, modified with the clearness indices KT_{min} and KT_{max} as seen above. The ESRA relationship can be written in the following generic form:

$$E(\lambda) = \left[c + d\lambda\right] f(\lambda) E_{pyranometer} \qquad (10.16)$$

Some simple calculations yield:

$$E(\lambda) = (c + d\lambda) E_{pyranometer} \left[KT_{max}\left(1 - f_{cloud}(\lambda)\right) - KT_{min}\left(1 - f_{clear}(\lambda)\right) \right. \\ \left. + KT\left(f_{cloud}(\lambda) - f_{clear}(\lambda)\right)\right] / \left(KT_{max} - KT_{min}\right) \qquad (10.17)$$

Similar to the ESRA relationship, the above equation may be used to compute the irradiance over a given spectral range between 310 and 1000 nm. For example, I have obtained the following formula[3] applicable worldwide for the UV-A spectral range [315, 400] nm, by setting KT_{max} to 0.7 and by constraining KT in the interval [0.1, 0.7]:

$$E_{UV-A}(\lambda) = \left[7.210 - 2.365 KT\right] E_{pyranometer} / 100 \qquad (10.18)$$

10.5.3 Estimate using other meteorological variables

Many empirical models have been published using input variables other than those already seen. I will not detail them; I have chosen a few to illustrate. There are, for example, models based on the difference between minimum T_{min} and maximum T_{max} of air temperature at surface, like that of Bristow and Campbell:[4]

---

3 Aculinin A., Brogniez C., Bengulescu M., Gillotay D., Auriol F., Wald L., 2016. Assessment of several empirical relationships for deriving daily means of UV-A irradiance from Meteosat–based estimates of the total irradiance. *Remote Sensing*, 8, 537, doi:10.3390/rs8070537.

4 Bristow K. L., Campbell G. S., 1984. On the relationship between incoming solar radiation and daily maximum and minimum temperature. *Agricultural and Forest Meteorology*, 31, 159–166.

Table 10.1 Correction factors f_{clear} and f_{cloud} for clear-sky and cloudy conditions, respectively, for each wavelength λ

λ (nm)	f_{clear}	f_{cloud}	λ (nm)	f_{clear}	f_{cloud}	λ (nm)	f_{clear}	f_{cloud}
310	0.299131	0.052609	550	0.019879	0.000174	790	-0.110311	-0.147491
320	-0.072117	-0.436417	560	0.03685	0.03385	800	-0.10112	-0.144815
330	-0.470114	-1.083667	570	0.026988	0.048032	810	-0.084861	-0.114738
340	-0.16368	-0.507508	580	0.005614	0.025231	820	0.052705	0.122264
350	-0.045845	-0.320075	590	0.041526	0.100624	830	-0.0074	0.037108
360	0.104993	-0.13539	600	-0.01307	0.001933	840	-0.112113	-0.176035
370	0.094821	-0.128547	610	-0.004661	0.01879	850	-0.083078	-0.171275
380	0.193805	-0.002148	620	-0.006657	0.012116	860	-0.132166	-0.235954
390	0.227752	0.092124	630	0.015066	0.055732	870	-0.136049	-0.256693
400	-0.028009	-0.137366	640	-0.015301	0.022291	880	-0.138935	-0.241424
410	-0.03828	-0.170108	650	-0.000992	0.062014	890	-0.139798	-0.235224
420	0.032069	-0.072013	660	-0.02921	0.011542	900	0.115449	0.238556
430	0.187228	0.127635	670	-0.055281	-0.016555	910	0.111406	0.248487
440	0.088508	0.028846	680	-0.060831	-0.011822	920	0.000457	0.106361
450	0.001133	-0.071541	690	0.058582	0.124584	930	0.341127	0.553565
460	0.05125	0.008413	700	-0.023561	0.066365	940	0.438448	0.689698
470	0.095797	0.048258	710	-0.073968	-0.03606	950	0.42551	0.666235
480	0.040447	-0.01389	720	0.096588	0.240797	960	0.285605	0.540856
490	0.067742	0.026979	730	0.036455	0.151632	970	-0.070392	0.057491
500	0.07053	0.049169	740	-0.071741	-0.110799	980	-0.214837	-0.061214
510	0.0493	0.027436	750	-0.10414	-0.196811	990	-0.443858	-0.451716
520	0.085463	0.07242	760	0.423049	0.294647	1000	-0.470468	-0.524536
530	0.032227	0.007355	770	-0.061262	-0.137052			
540	0.037216	0.026798	780	-0.137308	-0.235992			

$$E_{day} = E_{0day} K_{T_{max}} \left[1 - \exp\left[-a \left(T_{max} - T_{min}\right)_{day}^{b} \Big/ \left(T_{max} - T_{min}\right)_{month} \right] \right] \tag{10.19}$$

where the parameter a is positive. Parameters a and b are empirically determined. $(T_{max}-T_{min})_{day}$ is the temperature difference for the day and $(T_{max}-T_{min})_{month}$ is that for the month. The model of Heargraves and al.[5] has the following form:

$$E_{day} = E_{0day} a \sqrt{\left(T_{max} - T_{min}\right)_{day}} + b \tag{10.20}$$

Supit and Van Kappel[6] added the cloudiness Cn expressed in oktas as follows:

$$E_{day} = E_{0day} \left[a \sqrt{\left(T_{max} - T_{min}\right)_{day}} + b \sqrt{\left(1 - Cn/8\right)} + c \right] \tag{10.21}$$

More complex methods have been proposed more recently, using more sophisticated mathematical tools than the functions presented above, such as neural networks, with some meteorological variables as input.

5 Hargreaves G. L., Hargreaves G. H., Riley J. P., 1985. Irrigation water requirements for Senegal River basin. *Journal of Irrigation & Drainage Engineering*, 111, 265–275.
6 Supit I., Van Kappel R. R., 1998. A simple method to estimate global radiation. *Solar Energy*, 63, 147–160.

Chapter 11

Control of the plausibility of measurements

At a glance

The plausibility of each measurement must be checked. The control assesses if the measurement is probably correct and if it is not too large or too small compared to what can be expected under the conditions encountered. The manner of carrying out a plausibility check is the same whether for measurements of total or spectral radiation, of global radiation or its components, regardless of the integration time: intra-hourly, hourly, daily, monthly, or yearly.

Generally speaking, the plausibility check looks for anomalies, that is to say, values, behaviors, or variations in values, which do not correspond to expectations or to typical values. First, typical values must be established, such as the extraterrestrial radiation or the radiation received on the ground in cloudless conditions. Then, a search for anomaly is performed, which is often based on a calculation of the difference between the measurement and the expected or typical value. If this difference exceeds a threshold, then the result of the test is negative and the measurement is declared suspicious.

Measurements performed at a station are accompanied by metadata, which are data describing the station and the measurements. Metadata must be verified, in particular the geographic position of the station, the time system, the time stamp of the measurements, and the units of the measurements. It is recommended to collect other information about the station, which helps to adjust the parameters of the control procedures.

Many institutes or companies operate ground-based instruments for measuring solar radiation, or estimate radiation by other methods such as those seen in the previous chapter. Many of them make these measurements and estimates available via the Internet or other means. This chapter and the following are intended for the users of such data proposed by these third parties, who they trust a priori.

The measurements of an instrument are only of value if they are of quality. Measuring solar radiation is quite complex. Many factors come into play, linked to the instrument itself, its calibration and monitoring of the calibration, the quality of the time stamp and clock for sampling, the way it was installed, the electronics associated with it and its possible drifts and failures, the electrical and electromagnetic environments, connection cables, possible shading, and influences of air temperature, relative humidity, frost and meteors (rain, hail, snow), and dirt (sands and other dust, bird droppings), to name only the most frequently encountered. It is therefore very difficult to give a complete guide to control the quality of the measurements that can be obtained from a third party, since detailed information is very often lacking.

However, it is possible to check the plausibility of each measurement, i.e., to verify that the measurement is probably correct, that it is not too large or too small compared to what one can expect under the conditions encountered. This control of the plausibility is the subject of this chapter and the following.

Although I generally use the terms measurement and instrument, I do not actually distinguish in this chapter and the following between a measurement made by an instrument on the ground and an estimate obtained by a meteorological reanalysis or images from a spaceborne sensor. In fact, the techniques for checking the plausibility are the same whether it is measurements or estimates.

Reader, you object that you are using measurements from a renowned institute. You stress that each measurement has a quality indicator provided by the said institute. Is it enough to use these indicators? No, it is not. You should check the plausibility of each measurement by yourself. Without questioning the procedures put in place by the institutes to assess the quality of the data they make available to others, you should not overlook the possibility that you got it wrong in downloading the measurements or in their use. Plausibility tests will highlight these errors if there are any. Of course, it is necessary to use the quality indicators already provided to possibly exclude or not to exclude measurements before the plausibility assessment, or else to check, even adjust, the parameters of the models used for the plausibility check.

I first present the objective of the plausibility check. Then, I discuss the principles of the check that is based mostly on the comparisons between the measurements and expected values. I indicate how to compute the expected values. Finally, I discuss how to verify the position of the station, the time stamp of the measurements, and their units. These metadata must be controlled first before performing visual inspections and automatic tests that are presented in the following chapter.

11.1 Objective of the control of plausibility

Checking plausibility of measurements means questioning each measurement. Is this measurement plausible? Is it likely that it is a representation of reality? Is it close to the solar radiation that was actually received on the ground? Is it not too large or too small compared to what can be expected under the conditions encountered, taking into account the uncertainty associated with the category of the quality of the measurements (high, good, moderate)?

Answering these questions is not obvious. How can you decide that the measurement is plausible if you are not particularly familiar with the measuring station and its operation? To do this, the measurements must be inspected according to what is known about the station and its climate. Anomalies are searched for in the values and in the form of the measurement series.

Let me take an example to illustrate the need to verify the plausibility of measurements even those made by renowned institutes. The measurements originate from the database of the World Radiation Data Centre[1] (WRDC), which is the world data center sponsored by the World Meteorological Organization for solar radiation data. The database includes measurements of solar radiation and sunshine duration carried out at measuring stations in the meteorological networks. The WRDC performs itself a control of the plausibility of the measurements; a quality indicator is assigned to each measurement.

I downloaded the time series of daily irradiation H_{day} measured at the meteorological station in Addis Ababa, the capital of Ethiopia. The latitude of the station is 8.983°, and its longitude is 38.800°. The elevation is 2324 m above the mean sea level. The selected period ranges from 1988-01-01 to 1992-12-31. I downloaded only daily irradiations indicated as non-suspicious by the WRDC. The time series is drawn in Figure 11.1. Three periods are clearly visible in this figure. Many non-suspicious data are available during the first period, between 1988-01-01 and 1989-01-31. The second period is about 2 years long, covering 1989 and 1990, and many data are missing or declared suspicious. The third period ranges between 1991-02-01 and 1992-12-31. Many data are lacking also during the third period. Incidentally, this graph illustrates the difficulty of obtaining long times series of reliable measurements, as discussed in the previous chapter.

I have also charted in Figure 11.1 the extraterrestrial daily irradiation H_{0day}. Because the Addis Ababa station is located at latitude 8.983°, i.e., close to the equator, the yearly profile of H_{0day} exhibits an absolute minimum around the December solstice and a local minimum around the June solstice as well as two maxima around the equinoxes, as seen in Figure 3.5 (Chapter 3). Though I have already reported that sometimes the hourly and intra-hourly irradiations are greater than the extraterrestrial

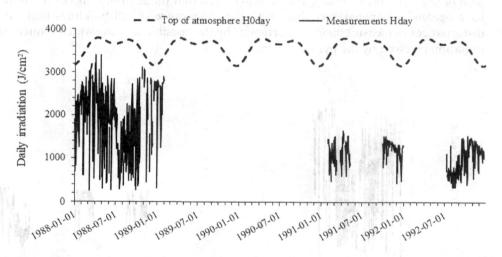

Figure 11.1 Daily irradiation H_{day} measured at the meteorological station in Addis Ababa, from 1988 to 1992. The dotted line is the extraterrestrial daily irradiation H_{0day}. (Source: World Radiation Data Centre (wrdc.mgo.rssi.ru) for H_{day} and SoDa Service (soda-pro.com) for H_{0day}.)

1 The database of the World Radiation Data Centre (WRDC) can be accessed at http://wrdc.mgo.rssi.ru/.

ones in specific cloudy conditions, this is not observed for daily values, and therefore, the daily irradiations H_{day} are expected to be less than H_{0day}. This is what can be seen in this graph, and this is a good point regarding plausibility.

Figure 11.2 exhibits the corresponding time series of the daily clearness index KT_{day}. During the first period covering 1988, the clearness index KT has a minimum of 0.07, an average of 0.55, and a maximum of 0.90. This average and this maximum are high. This is entirely plausible given that Addis Ababa is located on the Ethiopian high plateau, at an altitude of over 2000 m, and that the attenuation by the atmosphere is lower there than at sea level. The wet season is generally between April and September, which results in low daily irradiations as can be seen in Figure 11.1. From this discussion, one concludes that the measurements H_{day} of the first period are plausible. There is no anomaly or contradiction between the measurements and the expected typical values, given the knowledge of the station and the climate.

The same is not true for the third period (1991–1992), even if the measurements are declared valid in the WRDC database. It is obvious in Figure 11.1 that H_{day} is much less than that during the first period, approximately twice as small. The clearness index (Figure 11.2) has an average of 0.31 instead of 0.55 during the first period. One may note in Figure 11.2 that the clearness index seems to be clipped at approximately 0.4. The maximum is only 0.43, which corresponds to a cloudy sky, instead of 0.90 during the first period. Therefore, there are doubts as to the plausibility of these measurements. They are far too low compared to expectations and should be dismissed as suspicious. Put differently, they are likely not to be plausible because this behavior and these values cannot be explained, unless measurement problems are suspected. In the absence of details on the implementation of the instruments for this period, one can only express hypotheses, such as a defect in calibration or a decrease in the gain of the instrument. I have also seen firsthand that the surroundings of the station have become urbanized over time, with the appearance of tall buildings that cause disturbances in measurement, in particular by the possible shading which reduces the radiation received by the pyranometer.

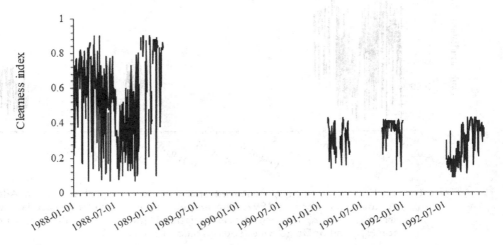

Figure 11.2 Daily clearness index at the meteorological station in Addis Ababa, from 1988 to 1992.

Reader, I may have wasted your precious time detailing this example. I wanted to because it illustrates the concept of plausibility check: establishing typical values and looking for anomalies. I also wanted to show the importance of knowing the station, its environment, and its climate, to get an idea of the expected values. Of course, no one can know all the places in the world, which constitutes a limit to the establishment of site-specific typical values. Fortunately, the Web, especially Wikipedia, can be of great help in establishing preliminary knowledge.

11.2 How to control the plausibility of measurements

The principles and manner of carrying out a plausibility check are the same whether it is for measurements of total or spectral radiation, of global radiation or its components, whatever the integration period: intra-hourly, hourly, daily, monthly, or annual.

In general, the control of plausibility consists of looking for anomalies, that is to say, values, behaviors, or variations in values, which do not correspond to expectations. Let me take a few examples. Value anomaly occurs when a measured daily irradiation is greater than the corresponding extraterrestrial one. Behavior anomaly occurs when the daily profile of hourly irradiation in clear-sky conditions does not exhibit a bell-shaped form, or if the monthly irradiation in winter is greater than that in summer in a temperate climate.

Looking for anomalies implies that typical values or bounds have been established, or even a model or a pattern of values that are expected, since an anomaly is an unexpected, unexplained deviation from standard values. Consequently, the plausibility check consists first of all in establishing typical values and limits and then looking for anomalies with respect to these values and bounds.

The purpose of the tests is to check the plausibility of the measurements. However, they do not guarantee reliably and every time that the measurement is correct or incorrect. For example, I wrote in Chapter 5 that, under certain cloudy conditions, it is possible to measure on a horizontal plane an irradiation greater than that at the top of the atmosphere for hourly or intra-hourly periods of integration. Tests point to such measurements as implausible when they are indeed correct. It is for this reason that I use the term *control of plausibility* and not *control of quality* in order to point out that a doubt may remain.

The search for possible anomalies is carried out by automatic procedures and visual inspections. But first check the description of the station and the information provided with the measurements. They help in establishing the typical values.

The measurements of a station are usually accompanied by metadata, more or less numerous and accurate. Metadata are data describing the station and the measurements, such as the geographic position of the station or the units of the measurements. The first thing to do is to check the metadata provided on the station. The second action is to collect information about the station, which is not included in the metadata, such as the climate, or its environment. It is also possible to consult the persons in charge of the station, or others who may have used the measurements of this station, and obtain their opinion.

Visual inspections and automatic tests are then performed to detect anomalies. They can be applied to radiation measurements, or to derivatives, such as the clearness index. Visual inspection involves an expert who will observe the data using various graphical tools, looking for anomalies. Extreme values or expected or typical values are optionally overlaid on the graphs to help detect unexpected values, behaviors, or variations in values. Examples of extremes are extraterrestrial irradiation or clear-sky irradiation, and examples of expected or typical values are monthly or annual climatological normals. In automatic procedures, the search for anomalies is often based on a calculation of the difference between the measurement and an expected typical value. If this difference exceeds a threshold, then the result of the test is negative and the measurement is declared suspicious.

There is a large variety of tests. Some tests include minimum and maximum typical values, that is, extreme values between which the measurements should lie. Other tests use rare typical values, that is, values rarely observed. Others make comparisons with the results of a clear-sky model. Others use other radiation observations, or even estimates of radiation obtained in another way, for example, by images of a spaceborne imager, or other meteorological measurements, by exploiting empirical relationships such as those seen in the previous chapter.

Visual and automatic tests can be chained. The order is according to the case. Tests can possibly be prioritized; that is, the results of some may be more important than others, depending on the operational objectives of the series of measurements. Typically, each test produces one result per measurement, which is an indicator of the plausibility of the measurement for that test. It can take several values. For example, an indicator equal to 0 means that the test is positive; that is, the measurement is plausible for this test. The value 1 may indicate that the result of the test is not sure, and the value 2 may indicate that the test is negative and that the measurement is not plausible. If several tests are done, they result in a set of plausibility indicators for each measurement. How to use these indicators to decide whether to keep or reject a measurement depends on each case and the objectives pursued. For example, if the goal is to have the longest possible series of measurements, the constraints may be relaxed and all the measurements that have successfully passed at least one or two tests are declared plausible. If the objective is to build a series of very reliable measurements which can serve as a reference in subsequent operations, then tests must be more draconian and success is requested at all the tests.

There are generally three types of tests for time series of meteorological measurements: tests on increments, tests on extremes, and consistency tests.

Tests on increments examine how the measurements change from moment to moment. As an illustration, let me take a series of hourly measurements of the air temperature at the surface, that is, 2 m above the ground. Test on increments verifies that the difference between two consecutive measurements is less than 8 °C in absolute value. For the same hourly step, the maximum differences are respectively 30 % and 15 m s^{-1} for the relative humidity at the surface (expressed in %) and the wind speed at surface (at 10 m height). The maximum differences depend on the integration duration. For example, they are equal to 3 °C, 10 %, and 20 m s^{-1}, respectively, for values integrated over 1 min or a few minutes. For such small integration durations, tests may also examine minimum differences over periods of 60 or 120 min, and the differences must be greater than 0.1 °C, 0.1 %, and 0.5 m s^{-1}, respectively, in absolute value.

There are no such upper or lower bounds on increments in irradiation due to the very nature of the solar radiation especially at timescales less than the day. As seen in Figure 6.2 exhibiting daily profiles of 15-min averages of irradiance, the appearance of a cloud over the measuring instrument induces a sudden drop in irradiation which may reach several hundred W m^{-2} and even more for the direct component as it will be seen in further examples. Fluctuations from moment to moment can be very large, and this prevents from setting effective upper and lower bounds. A maximum difference of 1000 W m^{-2} could be imposed on the irradiance, but this limit is so high that the test would be ineffective. In contrast, a test on the increments at the monthly scale can be effective if its parameters can be adjusted for the climate. For example, the monthly clearness index varies little from month to month at a site in a humid tropical climate like Mozambique or Malaysia (see, e.g., Figure 6.5), or in desert climate like in northern Africa or Australia. It varies more for a mid- to high-latitude site in a temperate climate like northern Europe or the west coast of North America. It turns out that tests on increments are rarely used for solar radiation unlike other meteorological variables. As a consequence, only tests on extremes and consistency tests are dealt with in this book. They are discussed in the following chapter.

Tests on extremes assess the plausibility of the smallest and greatest values. Consistency tests examine the consistency, or correlation, of different sets of measurements acquired by independent instruments. Both types of tests aim at determining whether the measurements lie within expected boundaries and limits and are therefore acceptable or not. These tests are guided either by physical reasoning to detect physically impossible events or by the statistical variability of the data to detect very rare and thus questionable events.

I will show in the next chapter that lower and upper bounds are often based on extraterrestrial irradiation H_0 or irradiation H_{clear} in cloudless conditions. Thus, performing a plausibility check requires being able to estimate such quantities. Models may help. For example, the equations in Chapter 3 are one way to calculate H_0. Many more or less complex models are available to estimate H_{clear}. The book by Daryl Myers documents several such models,[2] and several codes are available on the Web.[3]

The simplest models I know for timescales less than 1 day have only the solar zenithal angle θ_S as input. The model of Perrin de Brichambaut and Vauge[4] is as follows:

$$H_{clear} = 0.81H_0\left(\cos(\theta_S)\right)^{1.15} \tag{11.1}$$

A similar model is proposed in a technical note[5] of the World Meteorological Organization:

$$H_{clear} = 0.95H_0\left(\cos(\theta_S)\right)^2 \Big/ \left(\cos(\theta_S) + 0.2\right) \tag{11.2}$$

2 Myers D. R., 2017. *Solar Radiation: Practical Modeling for Renewable Energy Applications*, CRC Press, Boca Raton.

3 https://github.com/EDMANSolar/pcsol, written by Oscar Perpiñán Lamigueiro. [The author gives R codes for several tens of clear-sky models.]

4 Perrin de Brichambaut C., Vauge C., 1982. Le gisement solaire: évaluation de la ressource énergétique, Eds Technique et Documentation, Lavoisier, Paris, France. [Christian Perrin de Brichambaut (1928–1995) was a French meteorologist, former engineer at Météo-France, and former president of Météo et Climat (formerly the Meteorological Society of France) who worked extensively in favor of the popularization of science and was my mentor in solar radiation in the 1980s.]

5 World Meteorological Organization, 1981. Meteorological aspects of the utilization of solar radiation as an energy source. Technical Note no 172, WMO no 557, Geneva, Switzerland, 298 p. [The model is described on page 120, and its use is limited to elevations above the mean sea level less than 400 m.]

The overall accuracy of such models remains poor suffering from a lack of inputs.

Another practical means is to use online services delivering time series of H_{clear}. One example is the McClear application in the SoDa Service. It exploits the McClear clear-sky model and returns time series of H_{clear}. User-defined inputs are the place of interest or geographic coordinates, desired period of time, time system, and integration duration from 1 min to 1 month. The application itself reads the database of estimates of constituents of the cloudless atmosphere carried out daily by the European Copernicus Atmosphere Monitoring Service (CAMS). Being a Web processing service, the McClear application can be invoked automatically from a computer code written, for example, in Python or other language, thus offering an opportunity to fully automate the process.

In any case, reader, you must remember that both H_0 and H_{clear} result from models and therefore have some degrees of uncertainty. For example, while the estimate of total extraterrestrial radiation is fairly accurate, UV radiation fluctuates considerably from day to day and is not accurately estimated. The uncertainty on H_{clear} depends on the quality of the combination of the clear-sky model and its inputs. Reader, you must take these uncertainties into account when setting test limits and when analyzing results.

11.3 Checking metadata: an overview

Metadata are data about data. They describe the station and the measurements. They may be in numerical format such as the geographic coordinates or the indicator of quality supplied with each measurement, or textual such as the name of the instrument manufacturer.

The number and the quality of the metadata are highly variable. The geographic coordinates are usually included. Brand and type of the instrument are often given. The class of the instrument with respect to the World Radiometric Reference of the World Meteorological Organization is usually indicated in the Web site of the manufacturer. This provides the category of the quality of measurements as seen in Chapter 9: high, good, or medium or moderate. So a first idea of the uncertainty of the measurement may be obtained.

Usually, little information is available on the location of the station, the terrain, the surrounding soils, or on how the instruments were installed, their implementation, their calibration plan, their maintenance, and the control of their measurements. Given the importance of the geographic positions, the surrounding relief, the time stamp, and the units, the metadata provided by the site must be checked and possibly completed. A good plausibility check requires a good description of the station and the measurements. Is the station located at the geographic coordinates mentioned? Is there any shading that affects the radiation received? Are measurements actually in the units and time system mentioned? These are some of the questions that must be answered before proceeding. Answers are rarely obtained by automatic procedures, and the verification of metadata is often done manually.

The metadata should indicate how the missing measurements are coded. This is seldom the case and should be checked, for example, by plotting measurements, or quickly visually scanning measurements in the file. Meteorologists often use the value −999 or −99 to indicate missing measurements. Some operators mark NaN (not a number), while others use the character – or a space. Sometimes the quality indicator accompanying each measurement carries a particular value for missing data. It may also happen that the operator has replaced the missing measurement by an estimate (see the previous chapter). The quality indicator then carries a particular value indicating this replacement operation.

Information that is not included in metadata may be collected through other sources of information. You can find climate maps and other climate data on the Web, or apps that provide more or less detailed satellite views of any part of the world. However, not everything can be found on the Web, or some information may be out of date. Contacting the station manager or the users of these measurements is often very effective in getting to know the station better.

11.4 Checking metadata: geographic location of the station

The solar radiation received on the ground strongly depends on the geographic position of the place, in particular because of the dependence of the radiation outside the atmosphere on the latitude and the geographic variation of the constituents of the clear atmosphere and of the clouds. The geographic coordinates provided in the metadata should be carefully checked.

Sometimes the geographic coordinates provided are misinterpreted and the latitudes and longitudes are reversed. It is also possible that the minus sign disappeared during the download of the data or during a subsequent format conversion, in which case the station is in the southern hemisphere instead of the north and vice versa, or in the east instead of the west. A glance at a map usually allows this kind of error to be cleared up, except when the latitude and longitude values are very close to each other in the first case, or if the location is close to latitude 0° or longitude 0° in the second case.

Personally, I often use applications on the Web giving satellite images from anywhere on the globe to check the geographic coordinates of the station. I can thus visualize the place indicated in the metadata and verify that it matches my expectations. These apps often give the elevation above sea level, which I can compare to that given in the metadata. By exploring the surroundings of the station, I get an idea of the surrounding relief and the possible shading created by this relief, depending on the relative position of the sun.

Geographic coordinates can be given according to the ISO 19115 convention, in which latitudes Φ are counted positive in the northern hemisphere and negative in the southern hemisphere. Longitudes λ are counted positive east of the 0° meridian and negative west of this meridian. For example, the couple (latitude: −23°, longitude: −37°) describe a site located in the southern hemisphere west of the 0° meridian. Geographic coordinates can also be given by specifying the hemispheres, e.g., latitude: 23° S, longitude: 37° W.

Elevation is often given in m above the mean sea level. Latitudes and longitudes are often given in degree. They may be given in sexagesimal notation, for example, 52° 45′ 12″, or decimal one, 52.7533°. As a reminder, converting from a sexagesimal notation (52° 45′ 12″) to a decimal is done by first dividing the number of seconds by 60 (12/60 = 0.2). The result is added to the number of minutes; the whole is then divided by 60 (45.2/60 = 0.7533). The result is added to the number of degrees and forms the decimal part, i.e., 52.7533°. Reciprocally, converting an angle in decimal notation (52.7533°) to a sexagesimal one is done by first multiplying the decimal part by 60 (0.753 × 60). It gives the number of minutes in decimal format (45.198′). The decimal part is then multiplied by 60 (0.198 × 60), and this yields the number of seconds, i.e., 52° 45′ 12″.

11.5 Checking metadata: time system and time stamp

Three elements must be checked: the time system, the integration duration, and the time stamp with respect to the integration duration. It is also necessary to verify that these elements do not change during the period covered by the time series.

According to the World Meteorological Organization, the time system should be true solar time (TST) and the time-stamping should be done at the end of the integration duration. For example, for hourly irradiations, the time 13:00 indicates the end of the integration period that took place between 12:00 and 13:00. These recommendations are not followed by all networks, including members of this organization. You must be vigilant about these elements, which can be very difficult to control for intraday integration durations.

11.5.1 Overview of the most common problems

Not all operators use the TST system. The time system used for measurements varies between stations and operators. The following systems are generally found: TST, coordinated universal time (UTC), and legal or civil time, that is, the time used legally in a country. Chapter 1 indicates how to convert legal time to UTC, mean solar time, or TST. True solar time is the one that allows calculating the position of the sun in a given place at a given time.

Consider first the case of integration durations greater than or equal to 1 day. In this case, the TST does not matter and the mean solar time is sufficient that is simply linked to UTC by a function of longitude. A date is associated with each measurement. It can be expressed in mean solar time or in UTC. It does not matter, except for longitudes close to the date line that varies by country. For these longitudes, it is possible to have a wrong day if the time-stamping takes place in UTC at the very beginning of the day; the day indicated is then the previous day. It is therefore necessary to specify the time system.

Regarding integration duration, it is necessary to find, in the metadata or other information, indications on the way in which the sum (irradiation) or the average (irradiance) is calculated, in order to verify whether the series of measurements can help to fulfill the objective pursued. For example, in a series of monthly values, the minimum number of valid daily irradiations used to calculate the monthly irradiation must be known. In another example, the minimum number of valid hourly irradiations used to calculate the daily irradiation must be known, as well as how the sum is computed.

The time-related problems are more acute in the case of intraday integration durations because the position of the sun during the day plays an important role on the solar radiation received at the top of the atmosphere and at ground.

Problems are often related to a lack of information contained in the metadata or in the representation of time and interval. I deal here with the case of digital time series of measurements provided by third parties. I assume that the user's goal is to have these data processed automatically by a computer. This implies that the computer is able to read and understand all the information relating to the measurements in order to limit manual interventions in the processing.

Usually, at least in my experience at the time of writing this book, the data provider includes a date and time with each piece of data. But a few additional elements are often missing to know for sure the time stamp of each datum: What does the provided time correspond to? Is this the acquisition start time, the end time, or some other choice, such as the middle of the acquisition interval? The World Meteorological Organization did prescribe that the time stamp should indicate the end of integration, but this prescription is not necessarily followed by all meteorologists or other communities carrying out radiation measurements. And how can the integration duration be known for sure? By the difference between the data acquisition times, you might say, reader. Certainly, but that would be assuming that the integration duration is identical to the sampling time, which is not known a priori. If it were the case, how to deal with time series with non-consecutive values, i.e., with time series exhibiting gaps? More generally, how to handle time series with irregular sampling? You then have to turn to the data provider and ask for details.

At best, and always according to my experience, the data provider gives a description in the metadata or in the header of the measurement file. For example, it may be written: *The integration duration is 10 min, and the time indicated corresponds to the start of the measurement.* The user of the data then has all the necessary information. However, this information is not coded in a standardized way. It can also be written in an equivalent way: *The time stamp is at the start of the 10-min integration.* The information cannot therefore be identified and then understood in a standard way by a computer. Manual intervention is required.

Many difficulties would be avoided and considerable time savings would be made if data suppliers used the ISO representation of the interval. The representation of time interval in the ISO 8601 standard on dates and times is done this way. Each measurement is accompanied by two dates and times, or a date and time and a duration. Thus, each measurement is entirely specified in time, and it is not necessary to resort to metadata.

I briefly touched on interval encoding in Chapter 9. I detail it further here. A time interval is expressed by a start instant and by an end instant. According to the ISO standard, the two representations, start and then end, are given and separated by a slash [/]. Typical representation is *YYYY-MM-DDThh:mm:ss/YYYY-MM-DDThh:mm:ss*. For example:

- 2017-05-02T13:34/2017-05-02T13:35 means an interval starting on 2 May 2017 at 13:34 and finishing the same day at 13:35 without any mention of the time system;
- 2019-10/2021-07 is an interval starting in October 2019 and finishing in July 2021 without any mention of the first and last days.

An interval can also be represented by a start time and its duration, or by its duration and an end time. Let me first present how the duration is coded. The letter P precedes the expression for the duration. The typical format for the duration is of the form *PYYYY-MM-DDThh:mm:ss*. For example:

- P0001-02-15T12:25:21 is a duration of 1 year, 2 months, 15 days, 12 h, 25 min, and 21 s;
- P0030 is a duration of 30 years;
- P0000-10 is a duration of 10 months;
- P0000-00-05 is a duration of 5 days;
- PT00:10 is a duration of 10 min. Note the difference with the duration of 10 months: P0000-10, by the presence of the letter *T*.

An alternative representation for the duration is *PnnYnnMnnDTnnHnnMnnS*. It may be preferred for its brevity. For example:

- P1Y2M15DT12H25M21S is equivalent to P0001-02-15T12:25:21 as above;
- P30Y is a duration of 30 years and is equivalent to P0030;
- P10M is a duration of 10 months and is equivalent to P0000-10;
- P5D is a duration of 5 days and is equivalent to P0000-00-05;
- PT10M is a duration of 10 min and is equivalent to PT00:10:00;
- P2W is a duration of 2 weeks;
- PT1H30M is a duration of 1 h and 30 min and is equivalent to PT01-30.

As written above, an interval can also be represented by a start time and its duration and is written as *YYYY-MM-DDThh:mm:ss/PYYYY-MM-DDThh:mm:ss* or using the alternative representation *YYYY-MM-DDThh:mm:ss/PnnYnnMnnDTnnHnnMnnS*. The interval may also be represented by its duration and an end time. It is then written in the form *PYYYY-MM-DDThh:mm:ss/YYYY-MM-DDThh:mm:ss* or *PnnYnn-MnnDTnnHnnMnnS/YYYY-MM-DDThh:mm:ss*. Given an interval starting on 2 May 2017 at 13:34 and finishing 1 min later, it may be represented as follows:

- 2017-05-02T13:34/P0000-00-00T00:01 or
- 2017-05-02T13:34/PT1M or
- P0000-00-00T00:01/2017-05-02T13:35 or
- PT1M/2017-05-02T13:35 or
- 2017-05-02T13:34Z/PT1M or PT1M/2017-05-02T13:35Z where now there is a mention of the UTC (letter *Z*).

Thus, the use of this representation of the interval removes all doubts. Several programming languages support this ISO standard, such as *datetime* in the Python language. Some providers have adopted the ISO standard. For example, the so-called Radiation Service of the Copernicus Atmosphere Monitoring Service, including the McClear clear-sky model, provides a time stamp to each line of results in the form 2018-01-01T13:00:00.0/2018-01-01T14:00:00.0, for example, for the estimate of hourly irradiation for the interval starting on 1 January 2018 at 13:00 and finishing the same day at 14:00.

11.5.2 Some graphs for the visual inspection of the time-related problems

How to check the time system, time duration, and the time stamp when the integration duration is less than 1 day?

When the integration duration is less than 1 day, the time system used is very important. Graphics are often used to verify that the time system indicated is the one actually used. I illustrate this point with the example of a time series of hourly irradiations H_{hour} measured at the station in Quezon City, Philippines. Its latitude, longitude, and elevation above the mean sea level are respectively 14.63°, 121.02°, and 51 m. The measurements were downloaded from the WRDC database. According to the metadata, they are given in the TST system and the time stamp is done at the end of the hour. Thus, a measurement time-stamped at 12:00 is an integration from 11:00 to 12:00 (PT01H/12:00 or PT01/12:00 in the ISO 8601 standard on dates and times).

My first graph is made by superimposing all the daily profiles of hourly irradiation as in Figure 11.3a. In other words, I plotted H_{hour} for each hour from 00:00 to 24:00 for each day and I superimposed the plots for all days. I only used the July and August measurements in Figure 11.3; I could have done it with the measurements every day of the year. The upper envelope of these plots is made up of the greatest hourly irradiations found in the time series for each hour during the period considered. If these maxima are obtained under clear-sky conditions, the upper envelope corresponds approximately to a daily profile of hourly irradiation under clear-sky conditions for July and August. In Figure 11.3a, this envelope presents a bell-shaped profile. Its values increase as the time approaches solar noon (12:00 TST), and then decrease. On this graph, the maximum is reached for the 13:00 time stamp and corresponds to the moment when the sun is at its highest. The values are also high for the neighboring 12:00 and 14:00 time stamps. Note also that H_{hour} is greater than 0 between the 07:00 and 18:00 time stamps. These times correspond to sunrises and sunsets and are quite consistent with the TST system for that latitude in the northern hemisphere, near the

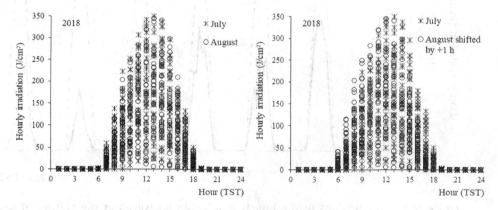

Figure 11.3 Superimposition of daily profiles of the hourly irradiation measured at the Quezon City station, in July and August 2018. (a) Actual measurements. (b) Actual measurements till 07-31 included, then artificial shift of +1 h. (Source: World Radiation Data Centre (wrdc.mgo.rssi.ru).)

equator. Thanks to this graph, I quickly verified that the time stamp used is compatible with the TST system as indicated in the metadata.

To show an example of an error on the time stamp, I faked a time stamp issue for the days in August. I artificially shifted the measurements by 1 h. This case of 1-h time shift is not purely theoretical; it is the one that is often encountered when time-stamping machines operate in civil time, or in legal time, and the country has adopted a summer and winter time system. Hence, the time series comprise correct measurements in July and shifted measurements in August. In other words, a time stamp of 12:00 corresponds to PT01H/12:00 in July and to PT01H/13:00 in August. I plotted the superimposed profiles in the figure but on the graph on the right. In this graph, the bell-shaped profile of the upper envelope is more crushed than previously and is no longer symmetrical. The greatest values are distributed between the timestamps 11:00 and 15:00, which is far from 12:00. The nonzero irradiations at the start and end of day are approximately at 06:00 and 07:00 on the one hand, and 17:00 and 18:00 on the other. The 06:00 and 17:00 time stamps are far from what can be calculated for sunrises and sunsets in TST which are close to 07:00 and 18:00 for this period. It is obvious that the circles are offset from the crosses. It can be difficult to express precisely the time stamp problem from this single graph. The sight of such a graph should arouse suspicion about the time stamp and lead to further analysis.

11.5.3 Benefit in using the hourly extraterrestrial irradiation

My second graph is a classical plot of the time series of the hourly irradiation throughout the year. For the sake of legibility, I have limited the plot in Figure 11.4 to the period from 12 to 18 January 2018 for the same station in Quezon City. To show an

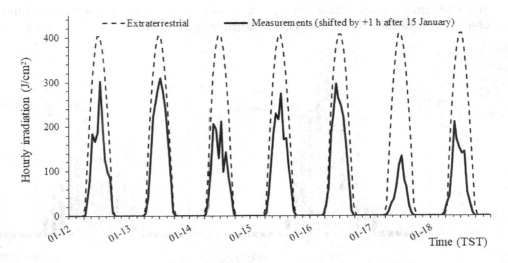

Figure 11.4 Daily profiles of the hourly irradiation measured at the station of Quezon City and that received at the top of the atmosphere, from 12 to 18 January 2018. Measurements have been artificially shifted by +1 h after 15 January. (Source: World Radiation Data Centre (wrdc.mgo.rssi.ru) for the measurements and SoDa Service (soda-pro.com) for the extraterrestrial irradiation.)

example of an error on the time stamp, I faked a time stamp issue after 15 January by artificially shifting the measurements by 1 h.

I also plotted the daily profile of the hourly irradiation H_{0hour} received at the top of the atmosphere. As expected, the daily profile of H_{hour} is included in that of H_{0hour} for the days from 12 to 15 January. Note that H_{hour} and H_{0hour} become nonzero and become zero again at the same time stamps, that is to say, that the sunrises and sunsets coincide between the two series. This is no longer the case after January 15 when the +1-h shift has been made. This is clear at the end of the days when H_{hour} is zero 1 h before H_{0hour}. There is therefore suspicion about the time stamp, and an analysis of the reason for the difference must be carried out.

Figure 11.4 shows that the maximum of H_{0hour} is reached at the 13:00 time stamp. It is difficult on this graph to determine the equivalent time stamp for H_{hour} because the sky conditions are cloudy as is common in this region. The daily profiles are rugged and H_{hour} does not necessarily reach its maximum for the smallest solar zenithal angle. This illustrates the advantage of the superposition of the daily profiles (Figure 11.3), which makes it possible to determine the time stamp of the maximum.

Compared to the superposition of daily profiles, Figure 11.4 uses additional information: the irradiation H_0 received at the top of the atmosphere. This irradiation must of course be calculated with the same time system and the same integration duration as the measurements. This additional information offers the possibility of verifying the other two elements: the integration duration and the time stamp with respect to the integration duration. In this example, the duration is 1 h. If this integration duration was wrong and had been less than or greater than 1 h, the profile of the measurements in Figure 11.4 would have shifted little by little compared to the profile of H_{0hour}. The profiles of H_{hour} would have been ahead of those of H_{0hour}, that is to say, more to the left on the plot if the duration had actually been less than 1 h. They would have been late if the duration had been longer than 1 h.

Thanks to the profiles of H_0, the fact that the time-stamping is done at the beginning, in the middle, or at the end of the integration duration is clearly visible. On the plot of Figure 11.4, the profiles of H_{hour} and H_{0hour} are in agreement between sunsets and sunrises, that is, when the sun is below the horizon. Since H_{0hour} was calculated with a time stamp at the end of the duration, this means that H_{hour} is also time-stamped at the end of the duration. If the time stamp had been done in the middle of the period, e.g., at 30 or 20 min, then there would have been a shift in the night profiles: H_{hour} would have been positive earlier than H_{0hour}, and conversely, H_{hour} would have become null sooner than H_{0hour}. In other words, the sun would have risen earlier and would have set earlier on the profile of H_{hour} compared to that of H_{0hour}. The same would have been the case for a time-stamping performed at the start of the integration duration.

Finally, it is desirable to check that the three elements – the time system used, the integration duration, and the time stamp – do not change during the period covered by the time series. This is done by plotting the same graphs as before for different sub-periods of the series, for example, the months of January, July, and December.

11.6 Checking metadata: checking the measured quantity and its unit

Metadata usually indicate the measured quantity and unit, but this information can sometimes be missing. In all cases, it is necessary to verify these elements and more precisely to verify whether the measurements are irradiations or irradiances, as well as the unit of this quantity.

To illustrate how to proceed, I am using in this section the hourly irradiations measured at the station in Alice Springs, Australia, during the year 2018. The latitude of this station is −23.80°, its longitude is 133.88°, and its altitude above the mean sea level is 547 m. The measurements were downloaded from the WRDC database and are given in the TST system. The time stamp is done at the end of the hour.

My goal being to give keys to recognize the measured quantity and the unit, I calculated different quantities from these measurements; then, I pretended not to know these quantities and their units. In my first example, I used the hourly values and plotted in Figure 11.5 the values for the 13:00 time stamp, i.e., from 12:00 to 13:00. Why choose this time stamp? Because it is easier to check the greatest values, and it is for this time stamp, or that of 12:00, that the greatest values are obtained if the sky is cloudless. I made two charts with the same measurements. The shapes are therefore the same. There is a minimum around June and a maximum around December, as the site is in the southern hemisphere.

On the left graph, the vertical axis shows values between 0 and 1250, while values range from 0 to 450 on the right graph. What is the quantity represented on the graph on the left? Hourly irradiation? Hourly average of irradiance? And what is the unit? Likewise, what is the quantity represented on the graph to the right? What is its unit? How to know these elements?

One way to estimate the measured quantity and its unit is to calculate the radiation at the top of the atmosphere, both the extraterrestrial irradiance E_0 and the irradiation H_0 and in several units (W m^{-2}, J m^{-2}, kJ m^{-2}, MJ m^{-2}, J cm^{-2}, Wh m^{-2}). The proximity of the resulting curves to those in Figure 11.5 is then evaluated and the quantity and its unit are deduced. By remembering that the clearness index in clear-sky conditions is about 0.7-0.8, and by multiplying E_0 and H_0 by this clearness index, a rough estimate of the solar radiation at ground in clear-sky conditions is obtained and the comparison with the curves in Figure 11.5 can be made. If a clear-sky model is

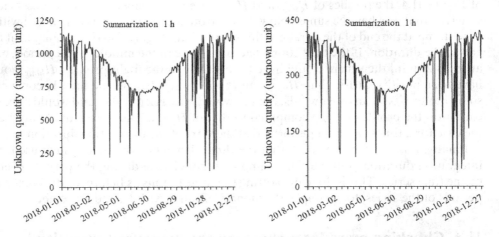

Figure 11.5 Example of checking the measured quantity and its unit based on hourly measurements performed at the station of Alice Springs, in Australia, for the time stamp 13:00. (Source: World Radiation Data Centre (wrdc.mgo. rssi.ru).)

available, the irradiance and irradiation curves can be calculated in several units and then one looks at which one is closest to each of the curves in Figure 11.5.

For my part, I quickly estimate the measured quantity and its unit from these curves by an arithmetic calculation. Knowing that the extraterrestrial irradiance E_{0N} at normal incidence is about 1360 W m^{-2}, I calculate the irradiance E_0 on the horizontal plane for a solar zenithal angle often encountered around solar noon in these latitudes, such as 30°. Since $E_0 = E_{0N} \cos(\theta_S)$, and since $\cos(30°)$ is approximately 0.9, I obtain an estimate of E_0 equal to 1220 W m^{-2}. I now multiply E_0 by the typical clearness index in clear-sky conditions, i.e., 0.8, to obtain an estimate of the clear-sky irradiance at ground which is approximately 980 W m^{-2}. This is very approximate, but it turns out that this value is close to the values taken by the curve for the left graph of Figure 11.5. So I can say that it is possible that the quantity represented is an hourly average of the irradiance expressed in W m^{-2}. I write "possible" because the integration time of 1 h is a special case, for which the hourly averages of irradiance are numerically equal to the hourly irradiations when the latter are expressed in Wh m^{-2}. It is therefore possible that the magnitude represented on the left graph is an hourly irradiation expressed in Wh m^{-2}. There is therefore an uncertainty that must be raised by collecting other information.

On the other hand, this value of 980 is much greater than the values of the graph on the right, which range from 0 to 450. This curve therefore does not represent irradiance but irradiation. An hourly average of irradiance of 980 W m^{-2} corresponds to an hourly irradiation of 3,528,000 J m^{-2} (multiplication by the number of seconds in 1 h, i.e., 3600 s), or approximately 3.53 MJ m^{-2} or 353 J cm^{-2} or 980 Wh m^{-2}. The values of the right graph of Figure 11.5 are of the order of 300. I can then affirm that the values represented on the right graph are hourly irradiations expressed in J cm^{-2}.

In the second example, I simulated measured values with an integration time of 10 min instead of 1 h. The way I did the simulation is of little importance for this illustration. As before, I pretended not to know the measured quantities or the units. I plotted in Figure 11.6 the values at time stamp 12:10 (integration between 12:00 and 12:10),

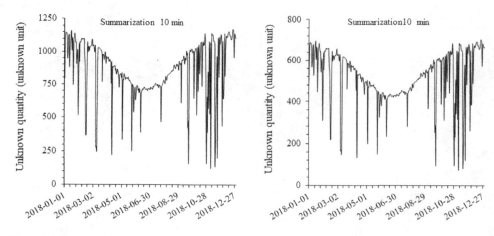

Figure 11.6 Example of checking the measured quantity and its unit based on 10-min measurements simulated from hourly irradiations measured at the station of Alice Springs, in Australia, for the time stamp 12:10. (Source: World Radiation Data Centre (wrdc.mgo.rssi.ru).)

i.e., the greatest values if the sky is cloudless (as well as for the time stamp 12:00). I made two charts with the same data.

On the left graph, the vertical axis shows values ranging from 0 to 1250, while values range from 0 to 450 on the right graph. I ask the same questions as before. What is the quantity represented on each of the graphs? An irradiation over 10 min? A 10-min average of irradiance? And what are the units? The previously made approximate calculation of irradiance under clear-sky conditions remains valid, i.e., 980 W m^{-2}. This value is close to the values taken by the curve on the left and greater than those of the graph on the right. I can then say that the quantity represented in the graph on the left is a 10-min average of irradiance expressed in W m^{-2}.

A 10-min average of irradiance of 980 W m^{-2} corresponds to a 10-min irradiation of 588,000 J m^{-2} (multiplication by the number of seconds in 10 min) or approximately 588 kJ m^{-2} or 0.59 MJ m^{-2} or 59 J cm^{-2} or 163 Wh m^{-2} (division by six, i.e., 10 min/60 min). The values of the right graph of Figure 11.6 are of the order of 600. I can then affirm that the values represented on the right graph are 10-min irradiations expressed in kJ m^{-2}.

Chapter 12

Visual and automated procedures

At a glance

The control of plausibility assesses if the radiation measurement is probably correct and if it is not too large or too small compared to what can be expected under the conditions encountered. Once the metadata have been verified, visual and automated procedures take place. Their concepts are the same whether for measurements of total or spectral radiation, of global radiation or its components, regardless of the integration time: intra-hourly, hourly, daily, monthly, or yearly.

In general, the plausibility check looks for anomalies, i.e., values, behaviors, or variations in values, which do not correspond to expectations or to typical values. Visual and automated procedures are ways to detect anomalies.

Visual analysis procedures use a variety of graphical tools. They consist of drawing graphs to examine the measurements, or derivatives such as the clearness index, in relation to other information. The expected or typical values are possibly superimposed on the graph, and anomalies are detected visually.

Automatic procedures generally make comparisons between measurements and extreme values that the measurements should not exceed. Other tests use rare typical values, i.e., values rarely observed. Others make comparisons with the results of a clear-sky model. If measurements of global radiation and its components have been acquired by three independent instruments, then the three sets of measurements can be checked against each other.

This chapter follows on from Chapter 11. The two chapters deal with checking the plausibility of measurements. Chapter 11 presents the objective of the control of plausibility and its principles. It details how to control the metadata that accompanies the measurements. Once the metadata are verified, visual and automated procedures take place. Their concepts are the same whether for measurements of total or spectral radiation, of global radiation or of its components, whatever the integration time: intra-hourly, hourly, daily, monthly, or annual. These procedures are described in this chapter. It is not intended to present all possible procedures. This chapter constitutes a sort of vade mecum, a set of simple rules and procedures that make it quite easy to verify the plausibility of measurements.

The control of plausibility assesses whether the radiation measurement is likely correct and whether it is neither too large nor too small compared to what can be expected under the conditions encountered. Once the metadata are verified, visual and automated procedures take place. Usually, the plausibility check looks for anomalies, i.e., values, behaviors, or variations in values, which do not correspond to the expectations or to the typical values as described in Chapter 11. Visual and automated procedures are means of detecting anomalies. I first detail some visual techniques highlighting any anomalies. Automatic procedures are also presented. They consist of one or more tests that the measurements must pass satisfactorily. Otherwise, the result of the test is negative and the measurement is declared suspicious. The tests result in plausibility indicators, which you will use at your convenience depending on your objectives.

Like Chapter 11, although I generally use the terms measurement and instrument, I do not distinguish in this chapter between a measurement made by an instrument on the ground and an estimate obtained by meteorological reanalysis or images from a space sensor. In fact, the techniques for checking plausibility are the same whether it is measurements or estimates.

12.1 The principle of visual inspection

It is recommended to carry out several visualizations of the measurements or their derivatives. Automatic procedures may be incomplete, in the sense that they may not reveal all anomalies. Visual examination or visual inspection by an expert eye is a contribution to the detection of values, behaviors, or variations in values, which do not correspond to expectations. I have shown in the previous pages several examples of visual examination and in particular the interest of plotting on the same graph the measurements and the corresponding extraterrestrial radiation.

The visual inspection of measurements consists of examining the measurements, or derivatives such as the clearness index, in relation to other information, such as extraterrestrial radiation or that in cloudless conditions. It is therefore necessary to have the means to obtain this extraterrestrial or cloudless total or spectral radiation, as the case may be. I find the site www.soda-pro.com very practical for this; these radiations can be downloaded there for different integration durations, at least for the total radiation at the time of writing.

The visual control of measurements, total or spectral irradiance or irradiation, or their derivatives, is carried out by drawing and examining graphs. The extreme or expected or typical values are possibly superimposed on the graph, and anomalies are detected visually. I have chosen to present visual control through illustration by presenting and commenting on graphs made with available measurements. For this, I use some of the most common graphs. I start with daily and monthly values and end with hourly and intra-hourly values.

12.2 Visual inspection of daily measurements

I present two examples in this section. The first one deals with daily irradiation measured by a pyranometer. The second one deals with daily irradiation measured in the UV-A band ([315, 400] nm).

12.2.1 Daily and monthly total irradiation

Three series of daily irradiations, G_{day}, D_{day}, and B_{Nday}, respectively, are shown in Figures 12.1–12.3 which illustrate my point well. The exact place of acquisition being confidential, you will know of this site, reader, only its latitude: −42° and that it is located at sea level in a temperate climate. Three independent instruments measured the hourly irradiation G_{hour}, its diffuse component D_{hour}, and the direct component at normal incidence B_{Nhour}, during the year 2015. When there were 24 plausible hourly

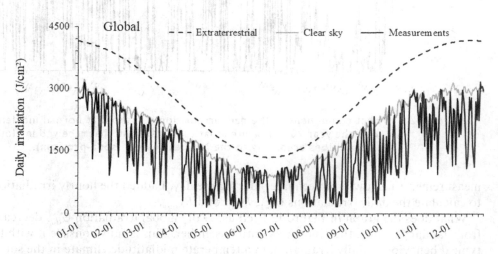

Figure 12.1 Daily irradiation measured during the year 2015 at a site at latitude −42° and that received in cloudless conditions according to the McClear model (soda-pro.com).

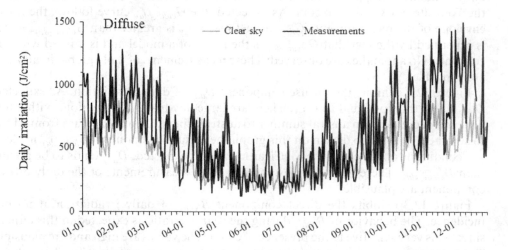

Figure 12.2 Diffuse component of the daily irradiation measured during the year 2015 at a site at latitude −42° and that received in cloudless conditions according to the McClear model (soda-pro.com).

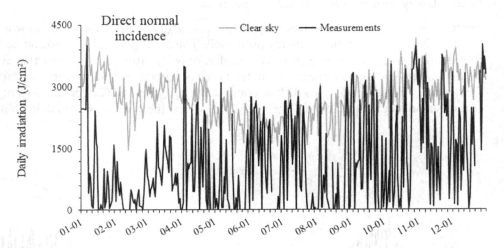

Figure 12.3 Direct component of the daily irradiation measured at normal incidence during the year 2015 at a site at latitude −42° and that received in cloudless conditions according to the McClear model (soda-pro.com).

measurements (as it will be seen later) during the day, I added the hourly irradiations to calculate the daily irradiations G_{day}, D_{day}, and B_{Nday}.

What does the graph in Figure 12.1 tell us? Daily global irradiation G_{day} decreases from January to July, then increases until December. This is fully consistent with the typical behavior of daily irradiation in a temperate midlatitude climate in the southern hemisphere. No behavioral anomalies are detected in this graph. Unsurprisingly, G_{day} is always smaller than the daily extraterrestrial irradiation G_{0day}. The greatest values of the clearness index K_T are around 0.8, an expected value. I also plotted the daily irradiation G_{clear_day} in clear-sky conditions obtained by the McClear model on the Web site www.soda-pro.com. As expected, the G_{clear_day} curve follows the upper envelope of the measurements G_{day}. Sometimes G_{day} is greater than G_{clear_day}, which is explained by the fact that G_{clear_day} is the result of a model and is tainted with uncertainty. No anomalies are observed: These measurements of daily global irradiation are plausible.

Figure 12.2 exhibits the diffuse component D_{day} of daily irradiation. As expected, D_{day} is less than or equal to G_{day}. There are large day-to-day fluctuations with a tendency to decrease from austral summer to austral winter, then to increase from winter to summer. I have also drawn on this graph the diffuse component D_{clear_day} in clear-sky conditions obtained by the McClear model. As expected, D_{day} tends to be greater than D_{clear_day}. There is no obvious anomaly: The measurements of the daily diffuse component are plausible.

Figure 12.3 exhibits the direct component B_{Nday} of daily irradiation at normal incidence. The behavior of B_{Nday} during the year is erratic as expected in this climate since B_N is very sensitive to the presence of cloud. The values are high and are plausible. The greatest values of the direct clearness index $K_{T_{direct}}$ are around 0.7 and are plausible. I also plotted the direct component B_{Nclear_day} for clear-sky conditions obtained by the McClear model. The B_{Nclear_day} curve follows the upper envelope of the measurements B_{Nday} as expected, which reinforces the plausibility of the measurements. On the other hand, the measurements B_{Nday} are very low between 2015-01-07 and 2015-04-04. Is this an anomaly? Is there a natural explanation or are they suspicious measurements? An exceptional occurrence of cloudy days during these 3 months could explain the large number of low or zero B_{Nday}. But this would go hand in hand with a drop in G_{day} during this period, a drop which is not observed (Figure 12.1). Indeed, G_{day} is often close to G_{clear_day}. This non-decay of G_{day} combined with the decay of B_{Nday} could be explained by an increase in the contribution of the diffuse component D_{day}, but this is not observed either in Figure 12.2. The measurements D_{day} are no greater during these 3 months of austral summer than during the 3 months from October to December. There is no convincing explanation for this drop in B_{Nday} between 01-07 and 04-04 other than a problem with measurements of B_{Nday}. It must be concluded that the measurements of direct component of the daily irradiation do not seem plausible for this period.

Since measurements of global irradiation G_{day} and its components are available, I carried out an additional check, called a consistency test. It consists in verifying that G_{day} is indeed equal to the sum of its components on the horizontal plane; that is, $G_{day} = B_{day} + D_{day}$, apart from measurement uncertainties. This control can only be done if G_{day}, B_{day}, and D_{day} are measured by independent instruments. In this test, I draw on the same figure, on the one hand, the G_{day} series and, on the other hand, the $(B_{day} + D_{day})$ series. If the measurements are consistent, the two curves must overlap. But how to obtain the direct component B_{day} on a horizontal plane when only the direct component at normal incidence B_{Nday} is available? For this, I was inspired by the calculation of the effective solar angles seen in Chapter 2. In practice, I wrote:

$$B_{day}(j) = B_{Nday}(j)\, B_{clear\_day}(j) / B_{Nclear\_day}(j), \quad \text{for each day } j \qquad (12.1)$$

The irradiations B_{clear_day} and B_{Nclear_day} are given at each day by the McClear clear-sky model. I could therefore calculate B_{day} and obtain the sum $(B_{day} + D_{day})$. Figure 12.4 exhibits the curves of G_{day} and of this sum. The two curves overlap well between 01-01 and 01-06 and between 04-05 and 12-31. The disagreement becomes very important between 01-07 and 04-04. This confirms that the measurements of B_{Nday} during this 3-month period are not only suspicious; they are obviously wrong.

In fact, the automatic sun tracking system the pyrheliometer was mounted on was faulty during this period. The pyrheliometer was not aimed accurately at the sun. Hence, readings of B_{Nhour} were erroneous and too low, giving B_{Nday} estimates that were too low. Since the ring shading system for measuring the diffuse radiation is independent of this tracking system, the estimates of the diffuse radiation were correct.

This example illustrates the interest of such graphs and confirms the importance of visual inspection. Of course, a systematic calculation of the sum $(B_{day} + D_{day})$ and its comparison with G_{day} would have made it possible to detect the problem. Additional calculations would then have been made to identify the origin, or origins, of the problem: measurements of the direct component or of the diffuse component, or both together? Here, I have shown that thanks to the graphs, the origin of the problem is quickly identified.

Similar graphs can be used to assess the plausibility of a time series of monthly irradiations G_{month} and possibly D_{month} or B_{Nmonth} or B_{month}. I recommend to plot on the same graph three irradiations: the measured one G_{month}, the extraterrestrial one

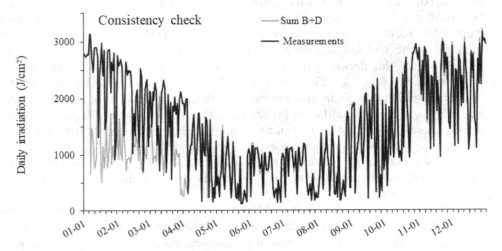

Figure 12.4 Verification of the plausibility of the measurements by superimposing the measured global irradiation and that resulting from the sum of the direct and diffuse $(B + D)$ components for 2015. The curves are superimposed when the measurements are consistent.

G_{0month}, and that in cloudless conditions G_{clear_month}. Similar to the daily case, G_{clear_month} is provided by a clear-sky model. G_{month} is expected to be smaller than G_{0month} and to be smaller than or close to G_{clear_month}. Otherwise, there is suspicion. The same is done with the direct component if it is available. B_{Nmonth} is expected to be smaller than $B_{0Nmonth}$ and to be smaller than or close to B_{Nclear_month}. Regarding the diffuse component, only D_{month} and D_{clear_month} are represented on the same graph since the diffuse component is zero at the top of the atmosphere. D_{month} is expected to be greater than or close to D_{clear_month}.

If measurements of G_{month} and D_{month} are both available and have been measured by two independent instruments, a consistency test can be carried out. It consists in plotting on the same graph G_{month} and D_{month}, and checking that D_{month} is always less than or equal to G_{month}. This test can also be performed with daily values.

If simultaneous measurements of G_{month}, D_{month}, and B_{Nmonth} were made using three independent instruments, another consistency test can be carried out, as seen above for the daily radiation. In this test, the measured global irradiation G_{month} and the sum of its two components $(D_{month} + B_{month})$ are plotted. If the measurements are consistent, the two curves must overlap.

If monthly normals are known for this site, or its vicinity, they can be plotted on the same graph as the monthly measurements. In this case, one expects that the latter are close to the former. Further visual inspection can include monthly averages and standard deviations of radiation and a comparison with what is known or expected. The clearness index K_T can also be calculated and plotted in relation to the expected values depending on the climate. A good indicator of plausibility is the histogram of the clearness index. For example, for a site in a temperate zone, such as in Europe or North America, K_T values less than 0.5 should be frequent, while for a Mediterranean type site where the number of cloudless days is high, K_T values greater than 0.5 should be common. Reader, you get it: You need to gather as much information as possible to make accurate comparisons.

12.2.2 An example of visual control of daily spectral values

The previous section deals with total daily radiation. The same process can be applied to radiation measured in narrower spectral bands. Usually, the operation is more complicated. Indeed, there are often less information on the extreme values than in the case of total radiation. At the time of writing, it is often difficult to estimate or obtain measurements of extraterrestrial radiation and solar radiation at ground under cloudless conditions in narrow spectral bands.

By way of example, Figure 12.5 exhibits a series of measurements of daily average of irradiance in the UV-A range ([315, 400] nm) covering the year 2010. Measurements were performed in Kishinev in Moldova, whose latitude and longitude are 47.00° and 28.82°, respectively, and elevation above the mean sea level is 205 m. Measurements were downloaded from the Web site woudc.org of the World Ozone and Ultraviolet Radiation Data Centre (WOUDC) of the World Meteorological Organization.

Figure 12.5 shows that overall, the daily average or irradiance E_{UVA_day} increases from January until June, then decreases until December. This is in agreement with the typical behavior of the daily average or irradiance E_{UVA_day} for a temperate mid-latitude climate in the northern hemisphere. The profile of E_{UVA_day} is very variable due to the great variability in cloud cover from day to day. High values of E_{UVA_day} compared to neighboring values can be noted for a few days from the end of January to the end of March and also in the last days of December. Are they too high? It is hard to say without additional information.

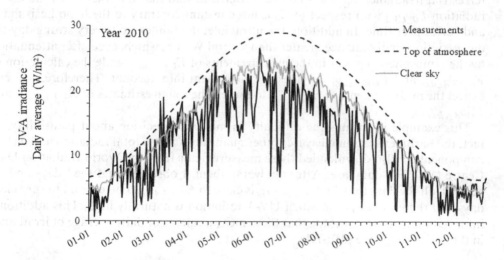

Figure 12.5 Daily average of irradiance in UV-A measured during 2010, at the station in Kishinev, as well as that at the top of the atmosphere and that received in cloud-free conditions according to the McClear model. (Source: World Ozone and Ultraviolet Radiation Data Centre (woudc.org) for measurements and McClear service (soda-pro.com) for extraterrestrial and clear-sky irradiations.)

In order to continue the visual inspection, is it possible to plot time series of extra-terrestrial irradiance $E_{0\_UVA\_day}$ and irradiance $E_{UVA\_clear\_day}$ received at ground in cloudless conditions as before? Yes, in principle this is possible. In practice, this would require that accurate and easy-to-use models be available, such as the McClear clear-sky model. At the time of writing, I do not know of any. It has been seen in Chapter 3 that the day-to-day variations in E_{0N} can reach 5 W m^{-2} and that the amplitude of the variations is 1–3 orders of magnitude greater in the UV domain than in the visible or the infrared range. It is difficult to know $E_{0\_UVA\_day}$ accurately and therefore to model it. However, it can be roughly estimated from the total extraterrestrial irradiance E_{0day}. The typical fraction of UV-A irradiance with respect to E_{0day} is about 6 % as seen in Table 7.2 of Chapter 7. If this ratio is assumed to be constant over time, multiplying the time series of E_{0day} by 0.06 yields a time series of estimates of $E_{0\_UVA\_day}$. Table 7.4 also gives the typical fraction of UV-A irradiance at ground $E_{UVA\_clear\_day}$ under cloudless conditions compared to the total irradiance E_{clear_day}. This fraction is about 7 %, and this ratio is also assumed to be constant over time. A clear-sky model such as McClear gives a time series of E_{clear_day} from which a time series of rough estimates of $E_{UVA\_clear\_day}$ is calculated by multiplying E_{clear_day} by 0.07. Both $E_{0\_UVA\_day}$ and $E_{UVA\_clear\_day}$ are plotted in Figure 12.5.

Figure 12.5 shows that, as expected, in general, E_{UVA_day} is smaller than the extra-terrestrial irradiance $E_{0\_UVA\_day}$ and smaller than or close to $E_{UVA\_clear\_day}$. Overall, the measurements appear plausible. But what about the high values mentioned above? There are 9 days for which E_{UVA_day} is greater than $E_{0\_UVA\_day}$. Should it be deduced from this that these measurements for these 9 days are not plausible? No, it is impossible to conclude. Indeed, the model estimating $E_{0\_UVA\_day}$ is too coarse for two reasons. First of all, it must be taken into account that the spectral distribution of the extra-terrestrial irradiance E_0 is not constant over time and that the fraction of the UV-A radiation E_{0_UVA} with respect to E_0 is not constant contrary to the hypothesis made and varies over time. In addition, the ultraviolet domain exhibits very strong day-to-day variations in irradiance greater than several W m^{-2}, which, even after attenuation by the atmosphere, appear in the measurements of E_{UVA_day}, while the calculation of $E_{0\_UVA\_day}$ and $E_{UVA\_clear\_day}$ does not take them into account. Therefore, one can expect the measurements of E_{UVA_day} to exceed the rough estimates of $E_{0\_UVA\_day}$ and $E_{UVA\_clear\_day}$.

This example illustrates the difficulty in making a decision about plausibility. In fact, the Kishinev station measures other quantities such as total radiation and its two components. I also downloaded these measurements from the World Radiation Data Centre (WRDC) database. After analyzing them, I observed that the 9 days and, in general, all the days for which E_{UVA_day} is close to $E_{0\_UVA\_day}$ correspond to very clear days and therefore days for which UV-A radiation is naturally high. This additional information helps to conclude that these measurements of daily average of irradiance in the UV-A range are plausible.

12.3 Visual inspection of hourly and intra-hourly values

Similar to daily or monthly values, visual inspection of hourly or intra-hourly values is carried out by analyzing graphs. I recommend drawing the same graphs as before, i.e., the time series of the measurements for the global radiation, or even the components if they are available. Corresponding values at the top of the atmosphere and in clear skies should also be plotted if possible. If series of measurements of global radiation and its components made by three independent instruments are available, then the consistency test of the three series is carried out with respect to each other.

The following graphs illustrate the visual inspection. As an example, I took hourly measurements made at the station of Sioux Falls, in the United States. The latitude and longitude of the station are respectively $43.73°$ and $-96.62°$; its elevation is $473\,m$ above the mean sea level. Three independent instruments measure the global radiation and its two components. I downloaded hourly irradiations from the WRDC Web site for the year 2018. For the sake of the legibility, I limited the plots to June 2018. I could have drawn for the whole year or more of course. The corresponding time series of extraterrestrial irradiance and irradiance in cloudless conditions were downloaded from the McClear clear-sky model on the Web site soda-pro.com. The three series of measurements G_{hour}, D_{hour}, and B_{Nhour}, respectively, are shown in Figures 12.6–12.8.

Figure 12.6 shows that the hourly global irradiation G_{hour} exhibits various profiles during the day. The daily maximum can vary a lot from day to day. For example, the first 6 days exhibit similar high maxima around $340\,J\,cm^{-2}$, but the next day, the maximum is only $150\,J\,cm^{-2}$. From 18 to 21 June, the sky was cloudy and the maxima are even less; it is only $50\,J\,cm^{-2}$ on 20 June. Such an alternation of clear and cloudy days is fully consistent with the typical behavior of hourly irradiation in a hot summer continental climate, i.e., typified by large seasonal temperature differences, with warm-to-hot and often humid summers and cold winters in the northern hemisphere. At Sioux Falls, the month with the most precipitation on average is June. Unsurprisingly, G_{hour} is always smaller than the hourly extraterrestrial irradiation G_{0hour}. As expected, the

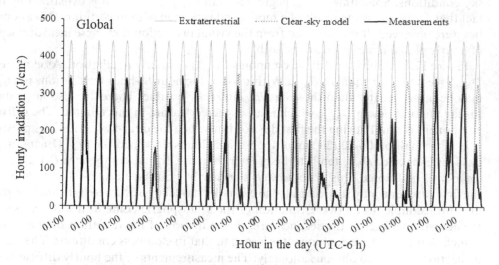

Figure 12.6 Hourly global irradiation measured during June 2018 at the station of Sioux Falls, in the United States, as well as that at the top of the atmosphere and that received in cloud-free conditions according to the McClear model. (Source: World Radiation Data Centre (wrdc.mgo.rssi.ru) for measurements and McClear service (soda-pro.com) for extraterrestrial and clear-sky irradiations.)

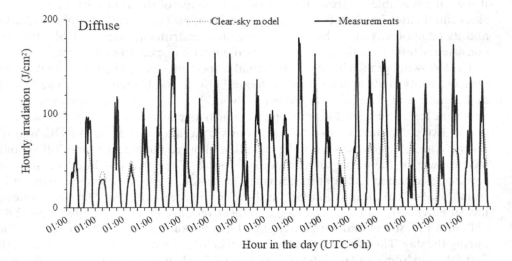

Figure 12.7 Hourly diffuse component measured during June 2018 at the station of Sioux Falls, in the United States, as well as that at the top of the atmosphere and that received in cloud-free conditions according to the McClear model. (Source: World Radiation Data Centre (wrdc.mgo.rssi. ru) for measurements and McClear service (soda-pro.com) for extraterrestrial and clear-sky irradiations.)

measurements G_{hour} are less than or close to the hourly irradiation G_{clear_hour} in clear-sky conditions. Sometimes G_{hour} is greater than G_{clear_hour}, which is explained by the fact that G_{clear_hour} is the result of a model and is tainted with uncertainty. No anomalies were observed. It is concluded from the visual inspection that these measurements of hourly global irradiation are plausible.

Figure 12.7 exhibits the diffuse component D_{hour} of hourly irradiation. As expected, D_{hour} is less than or equal to G_{hour}. Within a day, the hour-to-hour fluctuations in D_{hour} may be large, greater than those in G_{hour} in relative value. There are also large day-to-day fluctuations in D_{hour}, which is an expected feature in this climate. The diffuse component is small during the first days of June when the clear-sky conditions prevail. It is also small on 06-20 (20 June) when thick clouds darkened the sky. During most days, as expected, D_{hour} tends to be greater than the diffuse component D_{clear_hour} in clear-sky conditions. This is not always the case like on the days 06-03 and 06-20. In the first case, one may suspect that the clear-sky model overestimated the diffuse part due, for example, to an overestimation of the atmospheric content in water vapor or the aerosol loading. In the second case, the extinction of the radiation by the clouds is such that the diffuse component is less than that in cloudless conditions. The visual inspection reveals no obvious anomaly: The measurements of the hourly diffuse component are plausible.

Figure 12.8 exhibits the direct component B_{Nhour} of hourly irradiation at normal incidence. The behavior of B_{Nhour} from hour to hour and day to day is erratic as expected in this climate since B_N is very sensitive to the presence of cloud. The values are high and are plausible. The measurements of B_{Nhour} are less than or close to the estimates B_{Nclear_hour} in clear-sky conditions as expected, which reinforces the plausibility of the

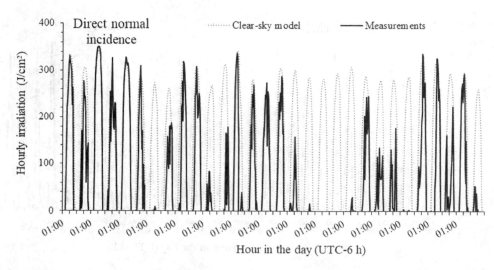

Figure 12.8 Hourly direct component at normal incidence measured during June 2018 at the station of Sioux Falls, in the United States, as well as that received in cloud-free conditions according to the McClear model. (Source: World Radiation Data Centre (wrdc.mgo.rssi.ru) for measurements and McClear service (soda-pro.com) for clear-sky irradiations.)

measurements. No anomalies were observed. It is concluded from the visual inspection that these measurements of the direct component are plausible.

Since measurements of global irradiation G_{hour} and its components are available and were measured by independent instruments, I carried out a consistency test. It consists in verifying that G_{hour} is equal to the sum of its components on the horizontal plane; that is, $G_{hour} = B_{hour} + D_{hour}$, apart from measurement uncertainties. How to obtain the direct component B_{hour} on a horizontal plane when only the direct component at normal incidence B_{Nhour} is available? Similarly to the previous case of daily irradiations, I was inspired by the calculation of the effective solar angles seen in Chapter 2. In practice, I wrote:

$$B_{hour}(t) = B_{Nhour}(t) B_{clear\_hour}(t) / B_{Nclear\_hour}(t), \quad \text{at each time } t \tag{12.2}$$

The irradiations B_{clear_hour} and B_{Nclear_hour} are given at each time by the McClear clear-sky model. I could therefore calculate B_{hour} and obtain the sum $(B_{hour} + D_{hour})$. Time series of the measured global irradiation G_{hour} and of the sum $(B_{hour} + D_{hour})$ are drawn in Figure 12.9. The two curves overlap well as expected; no anomaly is revealed. I conclude from this additional inspection that the three series of measurements are consistent and that all measurements are plausible.

12.4 Two-dimensional representation of hourly and intra-hourly values

I present in this section a new very useful additional chart, which is the two-dimensional representation (2D). It applies to radiation values or other derived quantities, such as

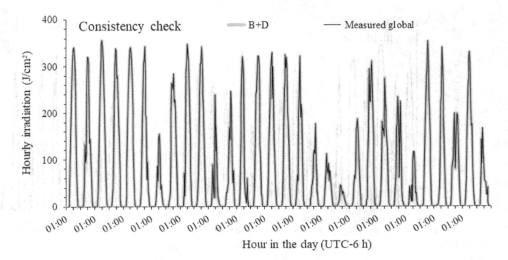

Figure 12.9 Verification of the plausibility of the measurements by superimposing the measured global irradiation and that resulting from the sum of the direct and diffuse (*B* + *D*) components for June 2018 at Sioux Falls, United States. The curves are superimposed when the measurements are consistent.

the clearness index. The horizontal axis is the day in the year and the vertical axis is the time in the day in this representation. In this way, a grid is obtained for the entire period covered, where each cell corresponds to a specific time in a specific day. Each cell receives a color, or, as here, a shade of gray, which corresponds to a class of radiation values. In other words, the radiation values are gathered into classes and a color is assigned to each class. A column in this grid corresponds to the daily profile of the radiation in a specific day. A line corresponds to the inter-daily profile of the radiation at a specific time. Such charts are also known as heat maps or heatmaps. They are used in many various domains.

Figure 12.10 illustrates this two-dimensional representation applied to the hourly global irradiations measured in 2018 at the Kishinev station. This station is located in a temperate zone, with generally clear-sky conditions often encountered between April and October. I recall the latitude, longitude, and elevation above the mean sea level of the station: 47.00°, 28.82°, and 205 m, respectively. Measurements were downloaded from the WRDC database. The days are plotted on the horizontal axis, from 01-01 to 12-31. The time of day is plotted on the vertical axis from 01:00 to 24:00. Irradiation is coded in grayscale, according to the legend to the right of the graph. Black is reserved for zero values, i.e., the times when the sun is below the horizon. The lighter the shade, the greater the irradiation. The white color combines irradiations greater than $200\,\mathrm{J\,cm^{-2}}$.

Note that the time on the vertical axis is the end time of the time stamp; i.e., 13:00 is an integration from 12:00 to 13:00 (12:00/13:00 in the ISO 8601 standard on dates and times, or PT01H/13:00 shortened to PT01/13:00).

What can be seen in Figure 12.10? The two-dimensional representation offers a synoptic view of the measurements. The white color appears especially when the sun is

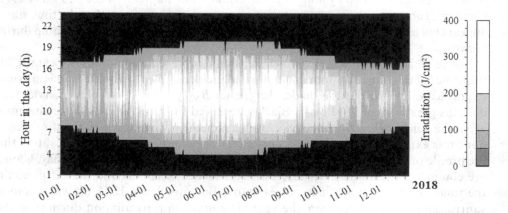

Figure 12.10 Two-dimensional representation of the hourly global irradiation measured at the Kishinev station in 2018. Black is reserved for null values. (Source: World Radiation Data Centre (wrdc.mgo.rssi.ru).)

high in the sky and between April and October, which is rather expected for this latitude and climate. Gray tones offer a gradient between white and black. This gradient confirms that the measured hourly irradiation increases from the moment the sun rises until solar noon, then decreases until sunset. The black shade represents the hours when the sun is below the horizon. The maximum irradiation during the day seems to be reached for the hour 12:00.

According to the metadata of the Kishinev station, the time stamps are made in the legal time system, or local standard, or civil, with no time change between winter and summer. The time-stamping is done at the end of the integration period as already said. In Kishinev, the legal time is UTC + 2 h. Is the time stamp correct? The two-dimensional representation makes it possible to answer this question. The first step is to calculate when the sun is highest in the three time systems: local time, UTC, and mean solar time (MST), as already seen in Chapter 1. Since the longitude of the station is 28.82°, the difference between MST at the station and UTC is 1.92 h (= 28.82°/15°). The MST is that for which the sun is highest at 12:00 on annual average. Therefore, 12:00 MST corresponds to 10.08 h (= 12.00−1.92) UTC or equivalently 10:05 UTC. True solar time, which is the time the sun is highest in the sky at 12:00 each day, fluctuates from −0.24 h to +0.28 h around MST during the year (see Figure 1.7 in Chapter 1). The sun is therefore highest between 9.84 h (= 10.08−0.24) and 10.36 h (= 10.08 + 0.28) UTC during the year or equivalently between 09:50 and 10:22 UTC. Add the 2-h shift from legal time in Kishinev, and you find that the maximum position is around 12:00 legal time. This is seen in Figure 12.10. So the time stamp seems correct.

Since the black shade indicates that the sun is below the horizon, the time interval between the two black areas along a vertical, that is, during the same day, defines the length of the day or daytime. This is fairly approximate since values are available every hour only. As expected, the daytime increases from winter (December solstice) to summer (June solstice) and decreases from summer to winter. The border of the lower black area indicates approximately the time of sunrise, while that of the upper black area indicates the time of sunset. These hours vary throughout the year, in accordance with the values obtained by applying the formulas in Chapter 2 (see, e.g., Figure 2.6 in Chapter 2) and taking into account the time stamp. This confirms that the time stamp appears correct. It can also be concluded that there is no shift in the time stamp during the year.

Figure 12.11 shows the same two-dimensional representation of hourly values, but this time for the direct component at normal incidence B_{Nhour}. Irradiation is encoded in grayscale. The lighter the shade, the greater B_{Nhour}. The white color includes irradiations greater than 250 J cm^{-2}. Black is reserved for zero values, i.e., for the hours when the sun is below the horizon as well as for the hours for which the direct component is extinguished by clouds. This global view of the measurements shows that the profile of B_{Nhour} fluctuates a lot during the day and also from day to day. Clouds are common in Kishinev. Because of this presence of clouds and the extinction of the time that they entail, it is not easy to follow in Figure 12.11 the variations in sunrise and sunset times over the year. The maximum irradiation during the day seems to be reached for the time 12:00, which is quite expected for a time stamp done in UTC + 2 h as discussed above. As with Figure 12.10, Figure 12.11 does not raise measurement issues.

Reader, you have noticed that in addition to the graphs seen in the previous sections, such two-dimensional representations offer a quick way to see possible problems around sunrise and sunset times, as well as time stamps. To illustrate this possibility, I introduced two artifacts into the measurements made at Kishinev and traced again the two-dimensional representation of the hourly global irradiation in Figure 12.12.

First, I simulated the presence of an important building northeast of the station. This building creates shading by intercepting the direct component of the radiation and attenuating the diffuse component. This is noticeable at the start of the day, around the June solstice when the solar azimuth at sunrise reaches its lowest values. This shading results in a decrease in radiation compared to other days and a small dark gray bump in Figure 12.12 in June. Conversely, reader, if you observe such a bump or a more

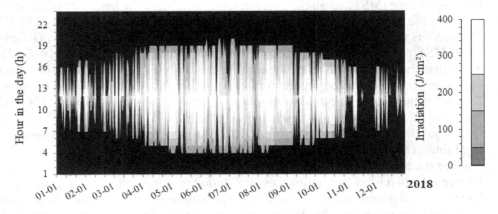

Figure 12.11 Two-dimensional representation of the direct component of the hourly irradiation at normal incidence, measured at the Kishinev station in 2018. Black is reserved for null values. (Source: World Radiation Data Centre (wrdc.mgo.rssi.ru).)

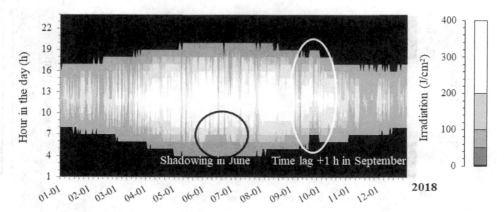

Figure 12.12 Two-dimensional representation of the hourly global irradiation measured at the Kishinev station in 2018 after introduction of artifacts. (Source: World Radiation Data Centre (wrdc.mgo.rssi.ru).)

complex shape at the beginning or end of the day and in general when the sun is low on the horizon, beware: It is possible that there is a shading of the instruments.

I also faked a time stamp issue for a few days in September. During these days, I artificially shifted the measurements by +1 h. The result appears clearly in Figure 12.12: The edges of the black areas are shifted upward by 1 h. Note that the irradiations, their values, and their forms are quite similar to those of the neighboring days: You can deduce that this is undoubtedly a simple shift of the time stamp of 1 h without influence on the measurements. This is indeed what I faked. In this case, everything is back to normal by simply modifying the time stamp during the days concerned without modifying the measurements.

Before closing this section, I also drew a two-dimensional representation of the hourly irradiation in a spectral band in Figure 12.13. My aim is to illustrate that these graphs are useful both for total radiation and for that in any spectral band, narrow or wide.

From the WRDC database, I downloaded hourly irradiation measurements made in 2015 by a pyranometer equipped with a so-called RG8 filter. The RG8 filter allows solar radiation to pass for wavelengths between 700 and 2900 nm. The station chosen is that of Bukit Kototabang, Indonesia. It is located on a plateau in a very large clearing surrounded by equatorial forest. The latitude of this station is $-0.20°$, its longitude is $100.32°$, and the elevation above the mean sea level is 864 m. Measurements are given in the local, or civil, time system, i.e., UTC + 7 h with no time change between winter and summer. The time stamp is done at the end of the hour.

As before, the days are plotted on the horizontal axis and the time of day is plotted on the vertical axis from 01:00 to 24:00. Irradiation is coded in grayscale as indicated by the legend on the right of the graph. Black is reserved for zero values, that is, the times when the sun is below the horizon. The lighter the shade, the greater the irradiation. The white color represents irradiations greater than 120 J cm^{-2}.

As expected, the white color appears especially when the sun is high in the sky. The gray gradient confirms that the measured hourly irradiation increases from the

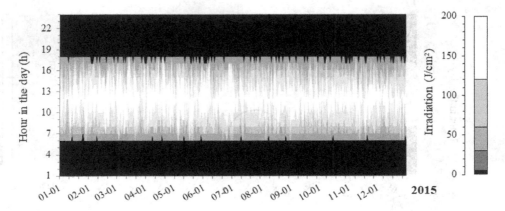

Figure 12.13 Two-dimensional representation of the hourly irradiation measured at the Bukit Kototabang station in 2015 in the spectral band [700, 2900] nm. Black is reserved for null values. (Source: World Radiation Data Centre (wrdc.mgo.rssi.ru).)

moment the sun rises until solar noon, then decreases until sunset. In this figure, the maximum irradiation during the day seems to be reached at 12:00 or 13:00. Is that correct? Given the longitude of the station (100.32°), the difference between the MST at the station and UTC is 6.69 h. Therefore, 12:00 MST corresponds to 5.31 h (= 12.00−6.69) UTC or equivalently 05:19 UTC. As seen in Figure 1.7 in Chapter 1, true solar time fluctuates from −0.24 to +0.28 h around MST during the year. The time of the sun highest is therefore between 5.07 h (= 5.31−0.24) and 5.59 h (= 5.31+0.28) UTC, or equivalently between 05:04 and 05:35 UTC, during the year. Knowing that the local time is UTC + 7 h, the time of the sun highest is between 12:04 (12.07 h) and 12:35 (12.59 h) local time during the year. The position of the maximum irradiation during the day in Figure 12.13 is therefore plausible, and the time stamp seems correct. This is confirmed by the times of sunrise and sunset, which are in line with expectations given the local time (see Figure 2.6 in Chapter 2). The edges of the black areas do not change over the year, which is normal for a site located at the equator. This means an absence of shading and an absence of drift in the time stamp during the year. The visual inspection of Figure 12.13 reveals no obvious anomaly with these measurements.

12.5 Automatic procedures

As with visual inspections, the goal of automatic procedures for checking plausibility of measurements is to detect anomalies if they exist. Anomaly finding is often based on a calculation of the difference between the measurement, or a derived quantity such as the clearness index, and an expected value. If this difference exceeds a threshold, then the test is negative and the measurement is suspicious. The great difficulty in automatic procedures is to find a compromise or trade-off on the constraints, i.e., on the expected values or the thresholds. A very restrictive test detects all suspicious measurements but usually creates false negatives, that is, non-suspicious measurements declared suspicious. If the constraints are relaxed for the test, i.e., if they are set so that the rejected values are really suspicious, the downside is that not all suspicious measurements will be detected and some will pass the test.

There are generally three types of tests for time series of meteorological measurements: tests on increments, tests on extremes, and consistency tests. As discussed in the previous chapter, tests on increments are rarely used for solar radiation. Therefore, this section does not present any. It focuses on tests on extremes and consistency tests.

Tests on extremes assess the plausibility of the smallest and greatest values. Consistency tests examine the consistency, or correlation, of different sets of measurements acquired by independent instruments. The principle is the same for these two types of tests: The quantity analyzed must be between two bounds. The limits must not be too small or high so as to have a discriminating power. As discussed above, too broad limits do not detect anomalies and too strict limits indicate too many measurements as non-plausible. There are generally two types of bounds. The rare bounds are minima and maxima rarely observed in measurements. The extreme bounds are minima and maxima which are not reached under normal conditions, or which must not be exceeded. They are sometimes referred to as physical limits. The range of extreme bounds includes the interval of rare bounds. If G, D, and B denote the global irradiation and its two components, typical forms of tests are:

$$\min \leq G \leq \max$$

$$\min \leq D/G \leq \max \tag{12.3}$$

$$\min \leq (D+B)/G \leq \max \tag{12.4}$$

Other types of bounds are possible such as the radiation values given by a clear-sky model. Performing tests with different bounds sets different constraints and can help in decision making about measurements. The limits are often fixed empirically. They are different depending on the integration time and possibly depending on the climate.

In the rest of this chapter, I describe some tests on extremes and consistency tests, often used for monthly, daily, hourly, or 1-min values.[1] I do not claim to be exhaustive since there are a multitude of similar tests that have been developed for different measurements, circumstances, and climates. I separated the tests for monthly and daily values from those for hourly and intra-hourly values because their shapes are slightly different. The tests detailed here apply to total radiation measurements or, at the very least, radiation measured in the broad spectral range of pyranometers. The case of measurements made in narrower spectral bands is discussed at the end of the chapter.

---

1 I draw inspiration from the bibliographical work performed during the European project ENDORSE (FP 7, 2007–2013), summarized in this publication: Espinar B., Blanc P., Wald L., Hoyer-Klick C., Schroedter-Homscheidt M., Wanderer T., 2012. On quality control procedures for solar radiation and meteorological measures, from subhourly to monthly average time periods. *Geophysical Research Abstracts*, 14, EGU General Assembly 2012, 22–27 April 2012, Wien, Austria. Available at hal-mines-paristech.archives-ouvertes.fr/hal-00691350.

12.5.1 Hourly and intra-hourly measurements

Solar radiation depends strongly on the solar zenithal angle at scales less than 1 day. It is therefore necessary to know the angle to be associated with each measurement, to perform the plausibility check of the hourly or intra-hourly values. If the integration duration is less than 1–2 min, this angle is the solar zenithal angle θ_S. Otherwise, calculate the effective solar zenithal angle θ_S^{eff} as seen in Chapter 2.

The tests published for hourly and intra-hourly measurements often include constants whose expression varies with the integration duration. This does not facilitate presentation of tests. For example, a constant irradiance of 100 W m^{-2} is equivalent to 360 kJ m^{-2} for hourly irradiation, 60 kJ m^{-2} for irradiation over 10 min, and 6 kJ m^{-2} for irradiation over 1 min. The equations presented here relate to the hourly or intra-hourly means of irradiance, and the constants are given in W m^{-2}. If necessary, it is up to you, reader, to carry out the appropriate conversions into irradiation.

Let E_0 and E_{0N} $\left(E_0 = E_{0N}\cos\left(\theta_S^{eff}\right)\right)$ be the irradiances received at the top of the atmosphere on a surface respectively horizontal and normal to the sun rays. For the global irradiance G (in W m^{-2}), the tests on the extreme limits and on the rare ones are written respectively:

$$0.03E_0 \leq G \leq \min\left[1.2E_{0N}, 1.5E_{0N}\cos\left(\theta_S^{eff}\right)^{1.2} + 100\right] \tag{12.5}$$

$$0.03E_0 \leq G \leq 1.2E_{0N}\cos\left(\theta_S^{eff}\right)^{1.2} + 50 \tag{12.6}$$

Similarly, for the diffuse component D (in W m^{-2}):

$$0.03E_0 \leq D \leq \min\left[0.8E_{0N}, 0.95E_{0N}\left(\cos\theta_S^{eff}\right)^{1.2} + 50\right] \tag{12.7}$$

$$0.03E_0 \leq D \leq 0.75E_{0N}\cos\left(\theta_S^{eff}\right)^{1.2} + 30 \tag{12.8}$$

and for the direct component B_N at normal incidence (in W m^{-2}):

$$0 \leq B_N \leq E_{0N} \tag{12.9}$$

$$0 \leq B_N \leq 0.95E_{0N}\cos\left(\theta_S^{eff}\right)^{0.2} + 10 \tag{12.10}$$

The equations above include a term of the form $\left(E_{0N}\cos\left(\theta_S^{eff}\right)\right)^{1.2}$ or $\left(E_{0N}\cos\left(\theta_S^{eff}\right)\right)^{0.2}$ for B_N, which is an approximate model of the solar radiation in clear-sky conditions as discussed in Chapter 6. It is possible to replace this term by another clear-sky model.

The consistency tests between the independent measurements of G and those of D, or between the measurements of G and those of B_N and D, can only be done if G is large enough, i.e., $G > 50$ W m^{-2}. Otherwise, the tests are not possible. For the two series G and D, the consistency test is:

$$D \leq 1.1G \tag{12.11}$$

In the case of the three series G, B_N, and D:

$$0.92 \leq \left(D + B_N \cos\left(\theta_S^{eff}\right)\right)\big/G \leq 1.08 \quad \text{if } \theta_S^{eff} \leq 75°$$

$$0.85 \leq \left(D + B_N \cos\left(\theta_S^{eff}\right)\right)\big/G \leq 1.15 \quad \text{if } \theta_S^{eff} > 75° \tag{12.12}$$

All the limits and constants given above must be adjusted according to the announced quality of the measurements (high, good, or moderate quality) and the associated uncertainties seen in Chapter 9. The values indicated here are those proposed by the high-quality network called Baseline Surface Radiation Network. The constraints must be relaxed if the measurements are of good quality, and even more so for the measurements of moderate quality in order to take account of the different levels of uncertainty.

12.5.2 Daily measurements

The daily global irradiation G_{day} should not exceed the corresponding irradiation H_{0day} at the top of the atmosphere. H_{0day} is often used to define the limits. The tests on the extreme limits and on the rare ones are written respectively:

$$0.01H_{0day} \leq G_{day} \leq b_{extreme\_day}H_{0day} \text{ with } b_{extreme\_day} \text{ between } 1.0 \text{ and } 1.2 \tag{12.13}$$

$$a_{rare\_day}H_{0day} \leq G_{day} \leq b_{rare\_day}H_{0day} \text{ with } a_{rare\_day} = 0.03 \text{ and } b_{rare\_day} = 0.9 \tag{12.14}$$

The parameter b_{rare_day} may be even set to 0.8 in conditions where the turbidity is often high in clear-sky conditions. Note that a_{rare_day} is the minimum of the daily clearness index and b_{rare_day} is its maximum. The same equations hold for the daily averages of irradiance E_{day} provided H_{0day} is replaced by the extraterrestrial daily averages of irradiance E_{0day}. This remark is valid for the following equations for the diffuse and direct components and for the consistency tests.

Another test may be performed where the upper rare bound is given by the irradiation G_{clear_day} calculated by a clear-sky model:

$$0.03H_{0day} \leq G_{day} \leq b_{rare\_day}G_{clear\_day} \tag{12.15}$$

where b_{rare_day} can be set to 1.1 if inputs to the clear-sky model are daily or intra-daily estimates of the atmospheric constituents. If inputs are rather climatological, then b_{rare_day} must be increased and can be set to 1.2 since the uncertainty on G_{clear_day} is greater in that case.

One may use the following tests for the daily diffuse component D_{day}:

$$0.03H_{0day} \leq D_{day} \leq 1.0H_{0day} \tag{12.16}$$

$$a_{rare\_day}H_{0day} \leq D_{day} \leq b_{rare\_day}H_{0day}, \text{ with } a_{rare\_day} = 0.03 \text{ and } b_{rare\_day} = 0.7 \tag{12.17}$$

where a_{rare_day} and b_{rare_day} are respectively the minimum and maximum of the daily diffuse clearness index.

As for the daily direct component B_{Nday}, it should be noted that B_{Nday} can be equal to zero. $b_{extreme\_day}$ and b_{rare_day} can be set to respectively 1.0 and 0.8, even 0.7 if the clear skies are often turbid. b_{rare_day} is the maximum of the daily direct clearness index. Tests have the following form:

$$0 \le B_{Nday} \le 1.0 H_{0Nday} \tag{12.18}$$

$$0 \le B_{Nday} \le 0.8 H_{0Nday} \tag{12.19}$$

What tests do you need to do on the direct component B_{day} received on a horizontal plane? Let θ_S^{eff} be the effective solar zenithal angle, i.e., the fictitious solar zenithal angle for the day, or month, that corresponds as best as possible to the measurement as seen in Chapter 2. It comes: $B_{day} = B_{Nday} \cos\left(\theta_S^{eff}\right)_{day}$, and $H_{0day} = H_{0Nday} \cos\left(\theta_S^{eff}\right)_{day}$. Transferring these values into the equations above shows that the parameters a and b are the same as for B_N and tests are:

$$0 \le B_{day} \le 1.0 H_{0day} \tag{12.20}$$

$$0 \le B_{day} \le 0.8 H_{0day} \tag{12.21}$$

If coincident and independent measurements of both G and D are available, a consistency test may be carried out of the form:

$$0.03 H_{0day} \le D_{day} \le 1.1 G_{day} \tag{12.22}$$

Sometimes G, D, and B_N are measured independently. A consistency test may be performed if G_{day} is large enough, typically greater than 1 MJ m^{-2} (or 12 W m^{-2}). It has the following form where $a = 0.1$:

$$(1-a) \le \left(D_{day} + B_{Nday} \cos\left(\theta_S^{eff}\right)_{day} \right) \Big/ G_{day} \le (1+a) \tag{12.23}$$

Similarly to the tests on hourly and intra-hourly measurements, all the limits given above must be adjusted according to the announced quality of the measurements. The values indicated here stand for good-quality measurements.

12.5.3 Monthly measurements

Similarly to the daily case, the monthly global irradiation G_{month} should not exceed the corresponding irradiation H_{0month} at the top of the atmosphere, the latter being often used to define the lower and upper bounds. The tests on the extreme limits and on the rare ones are written respectively:

$$0.05 H_{0month} \le G_{month} \le 1.0 H_{0month} \tag{12.24}$$

$$a_{rare\_month} H_{0month} \le G_{month} \le b_{rare\_month} H_{0month} \tag{12.25}$$

The parameters a_{rare_month} and b_{rare_month} are respectively the minimum and maximum of the monthly clearness index. The monthly radiation exhibits minima of clearness index which are greater than the daily minima and depend on the site in question. Thus, a_{rare_month} can be set to 0.5, or even 0.6, at a site with many occurrences of cloudless conditions. At sites in temperate zones, such as in Northern Europe, a_{rare_month} can be set to 0.2 and b_{rare_month} to 0.7.

The same equations hold for the monthly averages of irradiance E_{month} provided H_{0month} is replaced by the extraterrestrial monthly averages of irradiance E_{0month}. This remark is valid for the following equations for the diffuse and direct components and for the consistency tests.

Another test may be performed where the upper rare bound is given by the irradiation G_{clear_month} calculated by a clear-sky model:

$$0.03H_{0month} \leq G_{month} \leq b_{rare\_month}G_{clear\_month} \tag{12.26}$$

where b_{rare_month} can be set to 1.05 if inputs to the clear-sky model are daily or intra-daily estimates of the atmospheric constituents. If inputs are rather climatological, the uncertainty on G_{clear_month} is greater and b_{rare_month} must be increased and can be set to 1.1.

One may use the following tests for the monthly diffuse component D_{month}:

$$0.05H_{0month} \leq D_{month} \leq 1.0H_{0month} \tag{12.27}$$

$$a_{rare\_month}H_{0month} \leq D_{month} \leq b_{rare\_month}H_{0month}, \text{ with } a_{rare\_month}$$
$$= 0.1 \text{ and } b_{rare\_month} = 0.6 \tag{12.28}$$

where a_{rare_month} and b_{rare_month} are respectively the minimum and maximum of the monthly diffuse clearness index. Like for the global radiation G_{month}, a_{rare_month} and b_{rare_month} depend strongly on the geographical site. The rare upper bound b_{rare_month} can be set to 0.7 at a site with many occurrences of clear skies.

As for the monthly direct component B_{Nmonth}, it should be noted that B_{Nmonth} can be equal to zero. $b_{extreme\_month}$ and b_{rare_month} can be set to respectively 1.0 and 0.8, even 0.7 if the clear skies are often turbid. b_{rare_month} is the maximum of the monthly direct clearness index. Tests have the following form:

$$a_{extreme\_month}H_{0month} \leq B_{Nmonth} \leq b_{extreme\_month}H_{0Nmonth} \tag{12.29}$$

$$a_{rare\_month}H_{0month} \leq B_{Nmonth} \leq b_{rare\_month}H_{0Nmonth} \tag{12.30}$$

Both the extreme and rare bounds a_{month} and b_{month} are very variable depending on the geographic location for the monthly direct component B_{Nmonth} except $a_{extreme\_month}$ which is zero. At some sites, the direct component may be zero or almost zero during an entire month, while at others, the direct component is the main contributor to global irradiation throughout the year. It is therefore difficult to recommend universal values for $b_{extreme\_month}$ and for the two rare bounds. Otherwise, the same extreme limits as for daily radiation can be adopted. Like for daily values, the same tests may be adopted to check the plausibility of the direct component B_{month} received on a horizontal plane by replacing $H_{0Nmonth}$ by H_{0month}.

If coincident and independent measurements of both G and D are available, a consistency test may be carried out of the form:

$$0.05 H_{0month} \leq D_{month} \leq 1.1 G_{month} \tag{12.31}$$

In the cases where G, D, and B_N are measured independently, a consistency test may be performed if the monthly irradiation G_{month} is large enough, typically greater than approximately 30 MJ m^{-2} (or the monthly average of irradiance is greater than approximately 12 W m^{-2}). It has the following form where $a = 0.1$:

$$(1-a) \leq \left(D_{month} + B_{Nmonth} \cos\left(\theta_S^{eff}\right)_{month} \right) \Big/ G_{month} \leq (1+a) \tag{12.32}$$

As mentioned for the other integration durations, all the limits given above must be adjusted according to the announced quality of the measurements. The values indicated here stand for good-quality measurements.

12.5.4 Spectral radiation measurements

The automatic procedures presented above were designed for total radiation or, at the very least, radiation measured in the broad spectral range of pyranometers, ranging from approximately 330 to 2200 nm. What to do in the case of narrower spectral bands, such as the ultraviolet, or the visible?

Several documents detail how to perform measurements in various spectral bands, or the uncertainties of the corresponding instruments, or the quality controls recommended to the operators of these instruments. But, as far as the measurements received from third parties are concerned, to my knowledge, there are no plausibility check procedures that are approved by a community of users and as simple to implement as the previous procedures for total radiation.

While awaiting better times on this subject, I suggest using the same tests as for total radiation, but adapting them to the spectral band. The previous equations often refer to E_{0N} and E_0, which are calculated from E_{TSI}, which is the annual average of the extraterrestrial total irradiance at normal incidence. I suggest to proceed in the same way, but using $E_{0N}(\lambda_1, \lambda_2)$ and $E_0(\lambda_1, \lambda_2)$, calculated from $E_{TSI}(\lambda_1, \lambda_2)$, in these equations instead of E_{0N} and E_0. For practical implementation, a spectral distribution $E_{TSI}(\lambda)$ must be obtained in digital form. Several distributions $E_{TSI}(\lambda)$ are available on the Web. They are of sufficient quality to carry out an initial plausibility check though they may differ from each other. Given the desired interval $[\lambda_1, \lambda_2]$, the irradiance $E_{TSI}(\lambda_1, \lambda_2)$ is calculated by integrating the distribution $E_{TSI}(\lambda)$. If necessary, a spectral signature, or spectral response, is taken into account, such as the erythemal response, or the spectral luminous efficiency, or the response of a photovoltaic panel. Then, $E_{0N}(\lambda_1, \lambda_2)$ and $E_0(\lambda_1, \lambda_2)$ are deduced from $E_{TSI}(\lambda_1, \lambda_2)$, with the same equations as to calculate E_{0N} and E_0. Indeed, these equations, which are given in Chapter 3, do not depend on the wavelength. If necessary, the same approach can be applied to irradiations $H_{0N}(\lambda_1, \lambda_2)$, $H_0(\lambda_1, \lambda_2)$, and $H_{TSI}(\lambda_1, \lambda_2)$.

For illustration, let me take the example of the blue spectral band, of interval [480, 485] nm. According to Table 3.3 (Chapter 3), E_{TSI_blue} is 1 % of E_{TSI}, which implies that E_{0N_blue} and E_{0_blue} are respectively 0.01 E_{0N} and 0.01 E_0. The irradiances E_{0N_blue} and E_{0_blue} are transferred in the equations of plausibility tests instead of E_{0N} and E_0.

I draw your attention, reader, to the fact that these are fairly rough estimates, especially in the field of ultraviolet, subject to variations in irradiance at the top of the atmosphere, very strong from 1 day to another. So you have to be careful about their use.

Index

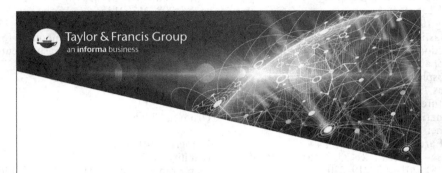

9780367725921